Elegant Table Linens

From Weil & Durrsè Including Wilendur™

Michelle Hayes

Schiffer® Publishing Ltd

4880 Lower Valley Road, Atglen, PA 19310 USA

Dedication

This book is dedicated to Bernadette Durrsè and Dorothy Durrsè Kinscherf for generously sharing the history of the Weil and Durrsè Company.

Acknowledgments

I thank Pamela Glasell, textile historian, for her encouragement in the writing of this book and Erin Henderson and Pamela Glasell for allowing me to photograph special Weil & Durrsè tablecloths from their private collections. I appreciate Judy Smasal allowing me to photograph my linens in her antiques store.

Library of Congress Catalog Control Number: 2004109761

Designed by Mark David Bowyer
Type set in Benguiat Bk BT/Aldine721 BT

ISBN: 0-7643-2090-4
Printed in China
1 2 3 4

Published by Schiffer Publishing Ltd.
4880 Lower Valley Road
Atglen, PA 19310
Phone: (610) 593-1777; Fax: (610) 593-2002
E-mail: Schifferbk@aol.com
For the largest selection of fine reference books on this and related subjects, please visit our web site at
www.schifferbooks.com
We are always looking for people to write books on new and related subjects. If you have an idea for a book, please contact us at the above address.

This book may be purchased from the publisher.
Include $3.95 for shipping. Please try your bookstore first.
You may write for a free catalog.

In Europe, Schiffer books are distributed by
Bushwood Books
6 Marksbury Ave. Kew Gardens
Surrey TW9 4JF England
Phone: 44 (0)20 8392-8585; Fax: 44 (0)20 8392-9876
E-mail: info@bushwoodbooks.co.uk
Free postage in the UK. Europe: air mail at cost.
Please try your bookstore first.

Contents

Introduction

Welcome to the world of Weil and Durrsè, makers of glorious household textiles. Those who fall in love with vintage linens are grateful that the Weil & Durrsè Company created artistic designs with brilliant rich coloring and exquisite shading on superior quality fabric. Their famous Wilendur brand is the benchmark for all others. The linens produced by Weil & Durrsè are the most sought after and desired of all the printed vintage household cloths. This was no accident. Weil & Durrsè stated on many of their labels, "Quality is never an accident. It is always the result of high intention, sincere effort, intelligent direction and skillful execution."

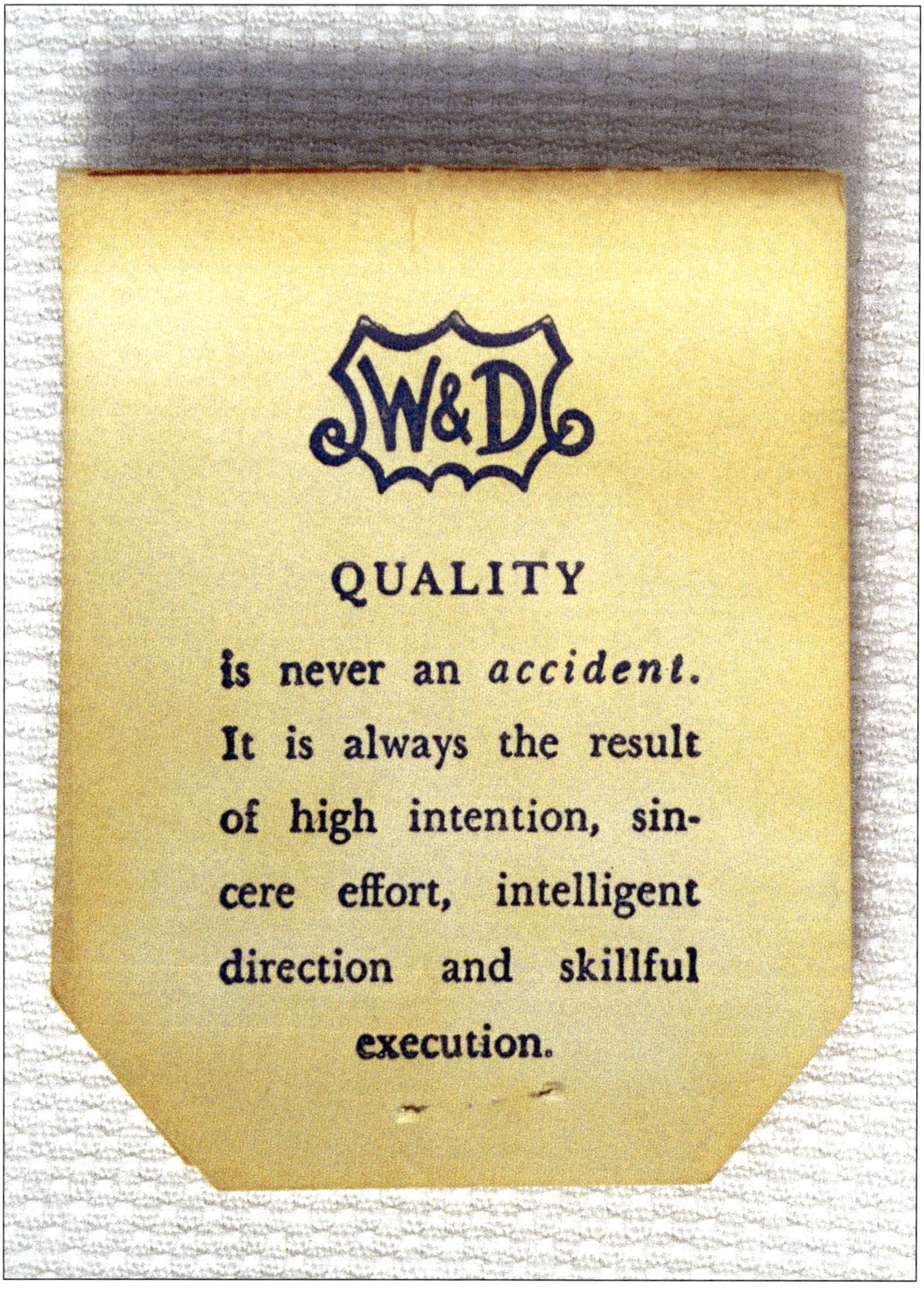

The Weil & Durrsè Company

The Weil and Durrsè Company was founded by Leopold Weil and Jacob F. Durrsè. Headquartered in New York City, it was located in the garment district. Jacob Durrsè began his career manufacturing silk goods. Toward the end of the First World War, Weil and Durrsè began producing household linens and in 1924 they introduced their Pride of Flanders line of linens. Durrsè made regular trips to Europe, in particular Belgium, where he purchased linen for the Pride of Flanders line. By the mid-1930s, the impending war in Europe made shipping linen to America difficult, and his last buying trip was in 1936, the year of the Olympic games in Berlin. At this point, the Weil & Durrsè Company began producing their goods from cotton sailcloth fabric, the fabric they are most well-known for today.

Leopold Weil and Jacob F. Durrsè each retired from the company in the 1950s and Jacob Durrsè's son, Jacob G. Durrsè, took over. He ran the company with his partners, John Goodman and Alex Guralnick, until both partners retired in the late 1970s. Then Jacob G. Durrsè ran the company alone until 1984, when he retired and closed the company doors for good.

Jacob F. Durrsè, the co-founder of Weil & Durrsè.

Group photograph of later owners: From the left, John Goodman, partner; Alex Guralnick, partner; Jack Halkin, manager of shipping; Jacob G. Durrsè, owner and son of Jacob F. Durrsè; on the far right, Max Steinkohl, whose wife, Shapsha, was the Weil & Durrsè bookkeeper.

Pricing and Condition

The value ranges listed with the captions to the photographs are based on actual prices paid for the items and an average range they commonly sell for. Today, vintage linens, especially the very popular Weil & Durrsè linens, have become increasingly popular and as desirability rises, so do prices.

When searching for vintage linens, do not let minor stains or tiny holes deter you from making a purchase. The obviously used cloths are the ones people have felt most comfortable enjoying and placing on their tables. Some stains can be lightened or removed (See Chapter 5.) and small holes can be darned or patched. Minor color fading can add a homey charm to a cloth, but I do not purchase badly faded linens. The value of a faded cloth is markedly lower than a cloth with good color.

Fabrics

The Weil & Durrsè Company used the following fabrics for their products:

Pride of Flanders™: Linen

Although Weil & Durrsè stopped producing this line for a period of time during World War II, they did resume the Pride of Flanders line after the war ended.

Wilendur™: Heavy cotton sailcloth

Wilendure™: Started with heavy cotton sailcloth

By the time they closed the business in 1984, Weil & Durrsè had used many fabrics in this line, including plastics, lightweight cotton, terrycloth, synthetic blends, and a sheer cloth with flocking.

Setting Pretty™: Heavy cotton

Oppa Tunity™: Heavy cotton
America's Pride™: Usually cotton/rayon blend
For a short period of time, this line was made of all cotton fabric.

For many years, Weil & Durrsè produced plastic tablecloths. They also produced terrycloth tablecloths that were very popular, but they stopped making them when polyesters were mixed with cotton. Many stores continued to ask for the terrycloth tablecloths because customers loved them.

Weil & Durrsè distributed their fabrics to large department stores to be sold off-the-bolts in varying lengths. The heavier-weight cloth was usually made into tablecloths, towels, and napkins. Curtains and valances were also made at home from this fabric. In the company's later years, their off-the-bolt cloth was made from a thin cotton fabric and a similar very lightweight polyester cotton blend, most often used for tablecloths and curtains.

Weil & Durrse also sold tablecloths to large department stores such as Sears, Roebuck and Company and J. C. Penneys to be sold under those stores own house labels such as Sears Harmony House and Penneys Fashion Manor.

Occasionally one will find a known Weil & Durrse design which will have another manufacturers label on it. These other companies were not affiliated with the Weil & Durrse company. Weil & Durrse patented only a small number of their designs so there were many designs which were open for copying by other manufacturers. On the patented designs the patent time frame was generally just 3 to 7 years so once that time frame ended those also were available for copying. There were several incidents where other manufacturers copied a Weil & Durrse design even before a patent had expired.

Weil & Durrsè used many different fabrics. Showing top to bottom, left to right, from earliest to most modern:
- linen
- heavy cotton
- cotton/rayon blend
- very light-weight cotton
- 95% linen, 5% rayon acetate
- medium-weight cotton with smooth finish (duracloth)
- 100% cotton terrycloth
- plastic with flannel backing
- light-weight cotton with slightly nubby texture
- 95% rayon, 5% acetate
- 50% polyester, 50% cotton

Product Types

Weil & Durrsè products include tablecloths, napkins, place mats, towels, furniture scarves and throws, curtains, draperies, bedspreads, toweling, table-cover fabrics, and cotton piece goods. These were sold in the U. S., Canada, Caribbean islands, and Africa.

Brands

The company produced six brands of products:

Pride of Flanders™
Wilendur™
Setting Pretty®
Wilendure®
Oppa Tunity™
America's Pride®

Pride of Flanders was the first line to be produced in 1924. Wilendur was first produced in 1938, America's Pride was first produced in 1941, and Setting Pretty was first produced in 1944. In 1958 Weil & Durrsè added an "e" to the name "Wilendur" making their new name, "Wilendure," in order to emphasize the *enduring* quality of their linens.

Colors

Mr. Durrsè had a talent in the field of color and design and was the primary force behind the decisions made on patterns and colors chosen for their linen goods. His talent is certainly evident in the spectacular artistry of the Weil & Durrse linens. Until the mid-1960s, the colors on Weil & Durrsè linens were exceptionally rich with deep shading and highlights. Typically, colors on the leaves in their patterns were unmistakable, because of their shading with light and dark green colors. They commonly used gray shadowing to accent their designs. These techniques: (1) rich colors, (2) highlights and shading of the leaves, and (3) gray shadows are signatures of a Weil & Durrsè cloth. These three characteristics identify their products and make them stand out from those of all other manufacturers.

From the mid-1960s forward, not only did their fabrics change, but also their colors were changing. As time went on, they stopped using the gray shadowing. Their colors were still beautiful, but were more subtle, not as brilliant. The highlighting techniques became muted and the leaves went from light and dark green shading to an olive green. Some leaves were also gray, soft green, and even black, often shaded with a touch of black, darker olive green, or white.

Although the fabrics and colors changed, Weil & Durrsè remained true to their company's policy statement, and the quality and artistic design of their linens remained exemplary.

Patterns

Today the most popular Weil & Durrsè tablecloths are those made in the 1940s and 1950s, specifically those of heavy cotton with bright colors and three-across patterns. The Weil & Durrsè Company did not produce more formal and expensive tablecloths because the founders' desire was to produce high quality cloths priced for the average customers.

The wives of the partners at one time suggested that the company should discontinue the block patterns that they were known for and produce border prints and other styles. This idea was discussed for several years because it was hard for the founders to abandon the use of the popular repeating patterns. Eventually, they did experiment with border prints and other styles and many were very successful for them.

Although the most well known Weil & Durrsè patterns today are the 3-across ones, they also made small floral squares covering the cloth, designs just around the sides, big bouquets in the corners, large florals in center only, simple stripes, tiny flowers all over, and more. All the different styles appear in the examples shown in this book.

Pattern and Fringe Examples

The most well-known Weil & Durrsè pattern is the "3-across" pattern, which shows a pattern in three large squares across the cloth. Three different types of the 3-across pattern follow.

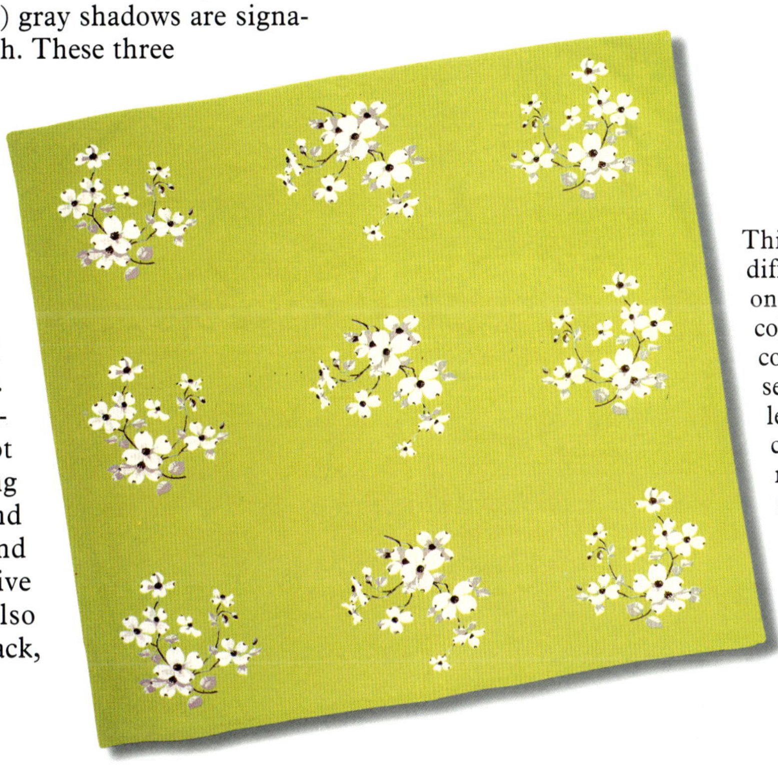

This cloth shows two different patterns – one in the center column and two side columns with a separate pattern. The left and right column patterns are mirror images. $20-$35

Color Variations

Also notice in the examples that follow different color variations of a same pattern. Along with their colored backgrounds, the Weil & Durrsè Company also used both bright white and cream backgrounds. Some of their cloths have a three-dimensional appearance because of the design's colors, while a similar cloth will have softer colors.

This cloth has three rows of the same pattern but the center column is rotated 90 degrees and the right column is rotated 180 degrees. $20-$35

Note the different fringes: the two on the left are the same, with the Christmas cloth having an added touch of gold trim. The fringe on the rose cloth is heavier and more elaborate.

This cloth shows two different patterns: one pattern in center column and another patttern on side columns.

Chapter 1
Tablecloths

Paper-tagged Tablecloths

Paper-tagged tablecloths are unused and still have their original paper tags attached. These paper tags give the design name, size, and name of the line the item was produced under (i.e. Wilendur, Pride of Flanders, etc.). Tagged items are still obtainable from flea markets, antiques shows, garage sales, and on-line auctions. Although many linens with the paper tags attached are shown on the following pages, the price ranges in the captions do not represent the ranges for paper-tagged items, unless specifically stated in the value line. Tagged items typically sell at significantly higher price ranges than untagged items. Although the beauty of the design is the most dependable way of pricing a paper-tagged item, the price range will vary for a variety of reasons. For example, values can vary when items are in an original box, when the box is in good condition and of good design, if napkins are included, if the size is large, and if the condition of the cloth is exceptionally good. Generally, tagged tablecloths sell between $30 and $110, with most in the upper end of this price range. Towels that retain their Weil & Durrsè paper stickers sell for about $10 more than those without a sticker.

Tablecloths are the most popular collector items from Weil & Durrsè. Over 250 different tablecloth designs are shown on the following pages and 120 Weil & Durrsè design names are listed. Weil & Durrsè often gave the same design name to two or even three different designs, and occasionally gave different names to different colors of the same design. Examples of this practice are shown throughout the book. Because Weil & Durrsè used many of the same designs in all of their fabric lines, you may be able to find the same floral design in, for example, Pride of Flanders linen, America's Pride cotton/rayon blend, and Wilendur cotton.

Paper tags

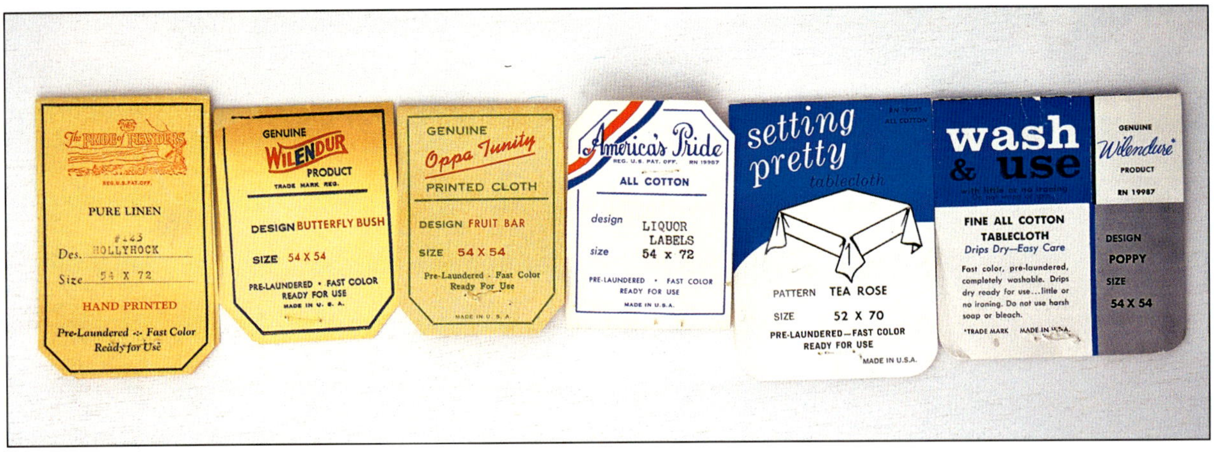

Showing, from left to right, oldest paper tags to most modern.

Showing the backs of the same tags.

Paper stickers on towels showing, from left to right, earliest labels to most modern.

Tablecloth Sizes

The Weil & Durrsè Company produced tablecloths in a number of different sizes. Their most common sizes are:

54" x 54"
54" x 72"
54" x 70"
48" x 54"
45" x 45"
36" x 36"
35" x 35"
34" x 34"
72" round
63" round

Of these, the most commonly sold size was 54" x 54", which would seat four people, followed closely by 54" x 72" seating four to six people, and 48" x 54" seating two to four people.

Large gatherings for weddings and church socials necessitated larger cloths, which Weil & Durrsè accommodated with tablecloths measuring 64" x 100" and 84" x 112".

Tablecloths were sold hemmed on all sides and were also sold with just two sides hemmed and two sides selvage. Tablecloths and towels were sold with and without sewn-in tags.

Meadow Bloom. Large floral clusters in each corner with oversized poppies and daisies. $20-35

Petunia. Glorious, richly colored petunias shaped in circles with connecting green stems. $35-50

Yellow daisies with a "loves me loves me not" look of scattered petals. $20-$35

Displays of large white poppies and multicolored daisies in each corner with a coordinating red boarder. $30-45

Clover. Pink and red clover with green multi shaded shamrocks. $30-45

Additional colors.

Oversized blue and white poppies with red centers, bordered by blue ropes. $20-35

Additional colors.

Mountain Ash. Mountain Ash berries in red and blue with green branches reaching from square to square. $30-45

Mountain Ash napkins. Set of four $20

Green ferns and yellow berries in a three-across design. $20-35

Fieldflowers. Stunning flower arrangements of red poppies, tulips, daisies and bachelor buttons. $20-$35

Large coral flowers on blue background with center of white leaves. $20-$35

Paper tag

Eye-opening, bright green background with little circles of jonquils. $20-$35

Additional colors

Pink and mauve roses with purple daisies and branches of white flowers. $20-$35

Paper tag

Rhododendron. Alternating gray and white squares showing the rhododendron flowers. Artist: John Madsen. Design date: 1947. $20-35

Red roses, tulips and more on colorful stripes. $25-40

Wildflower. Small individual flowers equally spaced cover this cheerful cloth. $20-35

Wildflower. Eight clusters of daisies and wheat on a white background with pink border. $20-35

Red poppies and blue Canterbury bells framed by a soft blue flowery border. $20-$35

Coffee Mill. Kitchen style tablecloth with old time coffee grinder, baskets of bread and twigs of bayberry. $20-$35

Red tulips and green leaves border this cloth. $20-35

Blue flowers and olive green leaves on a crisp white background. $20-35

Lilac Time. White and gray lilacs on a red background. Artist: Sergei Bogdanovich. Design Date: 1958. $30-45

Additional colors

Large bold red and blue nasturtiums with red border. $30-45

Additional colors

Yellow checked background with sunny yellow flowers. $20-$35

Bouquets of flowers with a swirl of gold and a blue border. $20-$35

Tropical red flowers with shaded green leaves. $30-45

Additional colors.

Black and orange floral design around sides. $18-25

African Daisy. Pretty pink and red daisies in a loose bouquet with drooping petals and seeds. $20-35

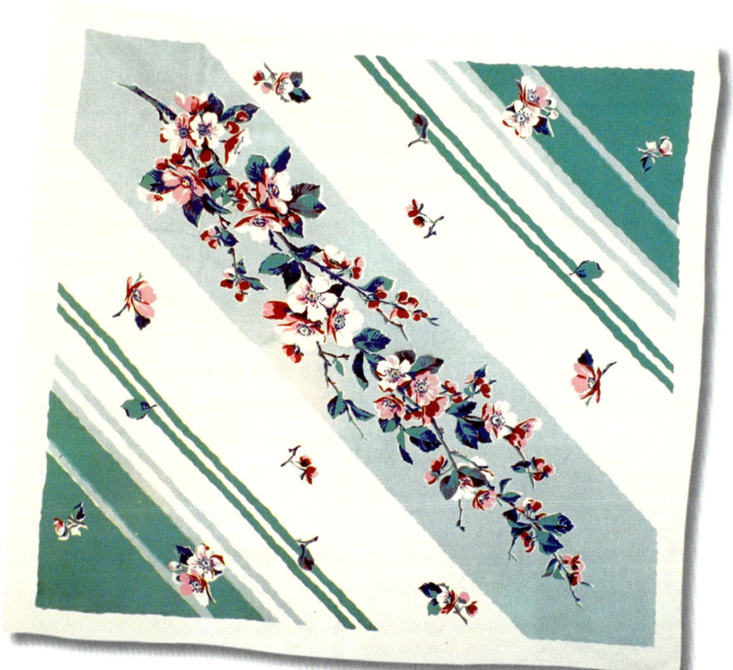

Pastel green and white background with pink and red blossoms through the middle. $20-35

Paper Tag on **African Daisy** in blue.

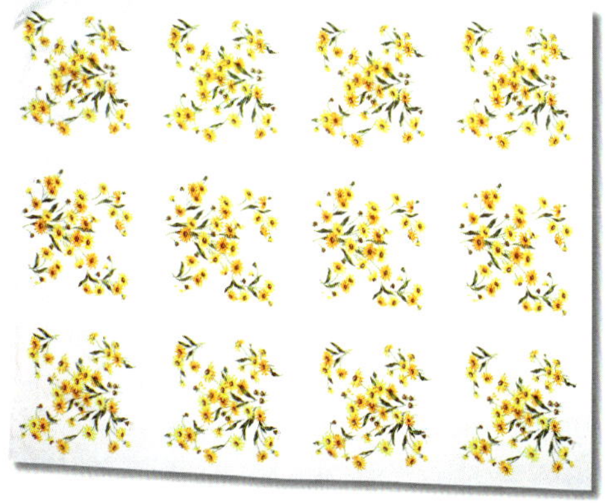

African Daisy in yellow.

Shasta Daisy. Yellow and white checkerboard design with squares of daisies. $20-$35

Squares of wrought iron style design in orange. $18-25

Red background with snowy white lily of the valley flowers. $20-$35

Kensington. Stylized brown tulips on brown and white 8 inch squares. $20-35

Paper tag

Red day lilies in leaf bordered squares. $20-35

Bloomberry. Cheerful yellow and red tulips set in gray squares. $20-35

Sunny yellow center with blue, green, yellow and red stripes forming a wide border all around. $20-35

Laurel. Unusual design showing white and gray background with red and blue floral squares. $30-45

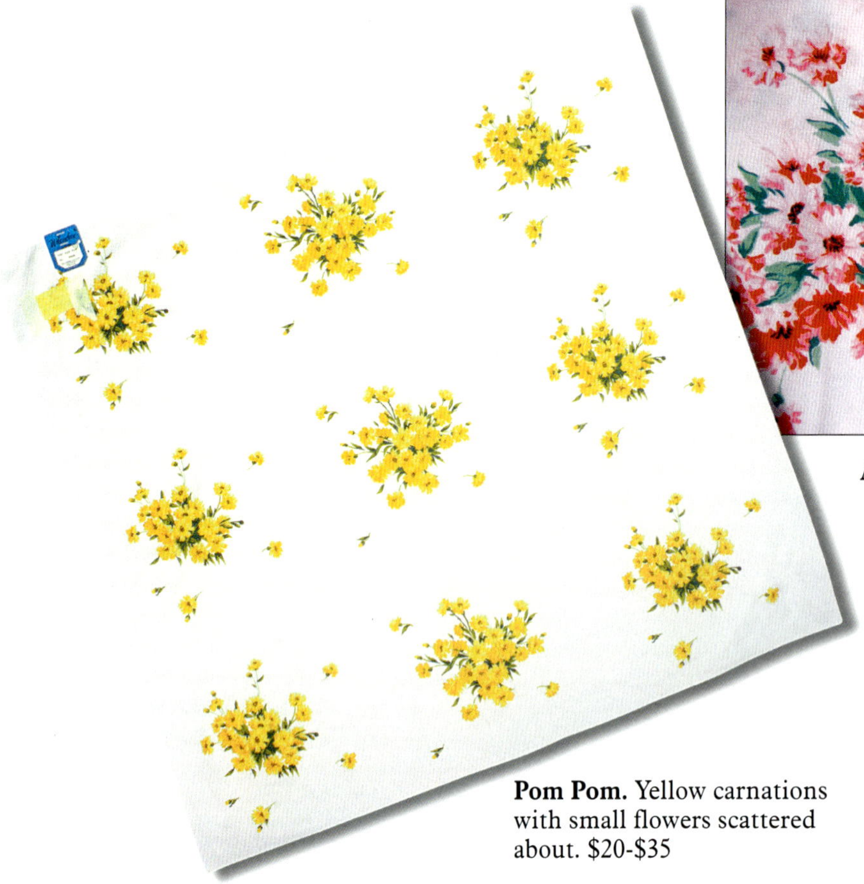

Additional colors

Pom Pom. Yellow carnations with small flowers scattered about. $20-$35

Pink and red dianthus with olive green stems scattered "willy nilly" around the border. $18-$25

Begonia. Strings of bright red flowers and green leaves with little flowers floating about. $30-45

Madison. Large gray and white squares showing gladiola in red, yellow and white. $20-$35

Floral and leaf design in burgundy and mauve. $20-35

Black-eyed Susan. Large yellow Black Eyed Susan flowers with rich red centers in a wide border around cloth. $30-45

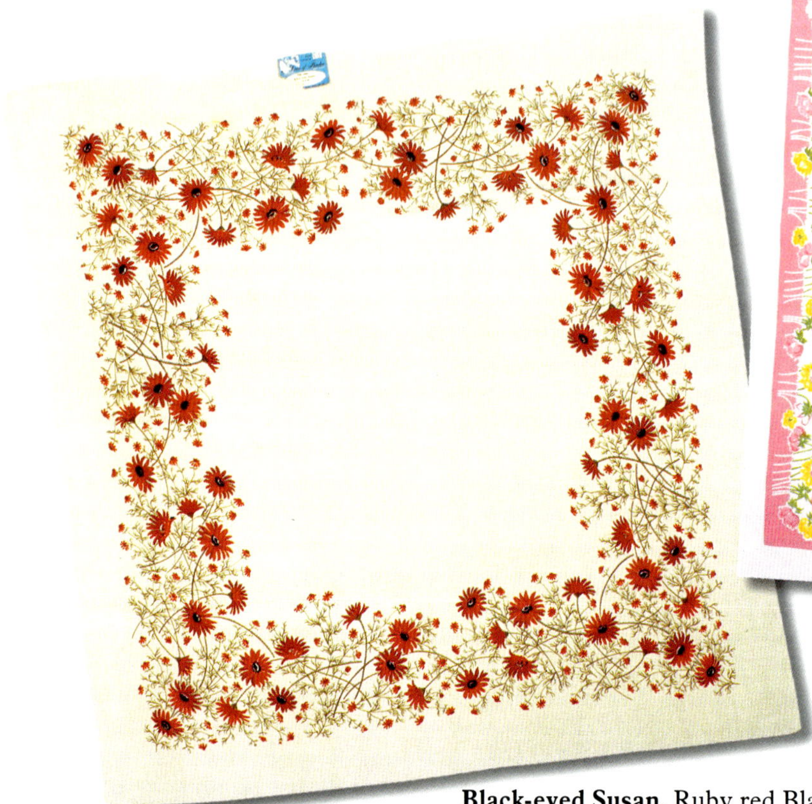

Anemone. Anemone, commonly called wind flowers, shown here in pastel shades of pink, yellow, and blue. $20-$35

Black-eyed Susan. Ruby red Black Eyed Susan flowers with lacey leaves. $20-35

Anemone. Big red anemone flowers with frilly olive green leaves. $20-$35

Paper tag

Red and lilac poppies with their soft fern like leaves. $20-35

Large purple zinnias on a blue and white background. $20-$35

Red flowers, lacey green leaves. $20-35

Paper tag

Hollyhock. Gorgeous large hollyhocks shown here on this brightly colored linen tablecloth.

Golden wheat sheaves on a rich red background. $20-35

White carnations on a bright green background. $20-$35

Matching napkins. Set of four $20

Windblown. Large sky blue flowers. $18-25

Carnation. Pink and red carnations. $20-$35

Paper tag

Oriental Poppy. Large yellow poppies with small gray flowers. $20-35

Maple Leaf. Maple leaves in pastel colors forming a wide border around cloth. $20-35

Hibiscus. Red hibiscus with soft green leaves. $20-$35

Lilac and blue with a border of yellow dogwoods. $20-$35

Sherbrooke. Flower and leaf "tree" growing out of each corner forming a wide border. $20-$35

Paper tag

Ivy. Clusters of twining green ivy in a 3 across pattern. Artist: Sergei Bogdanovich. Design date: 1956. $20-35

Paper tag. Note the original store price tag is still attached to cloth. This 45 x 45 inch tablecloth was priced at $1.98.

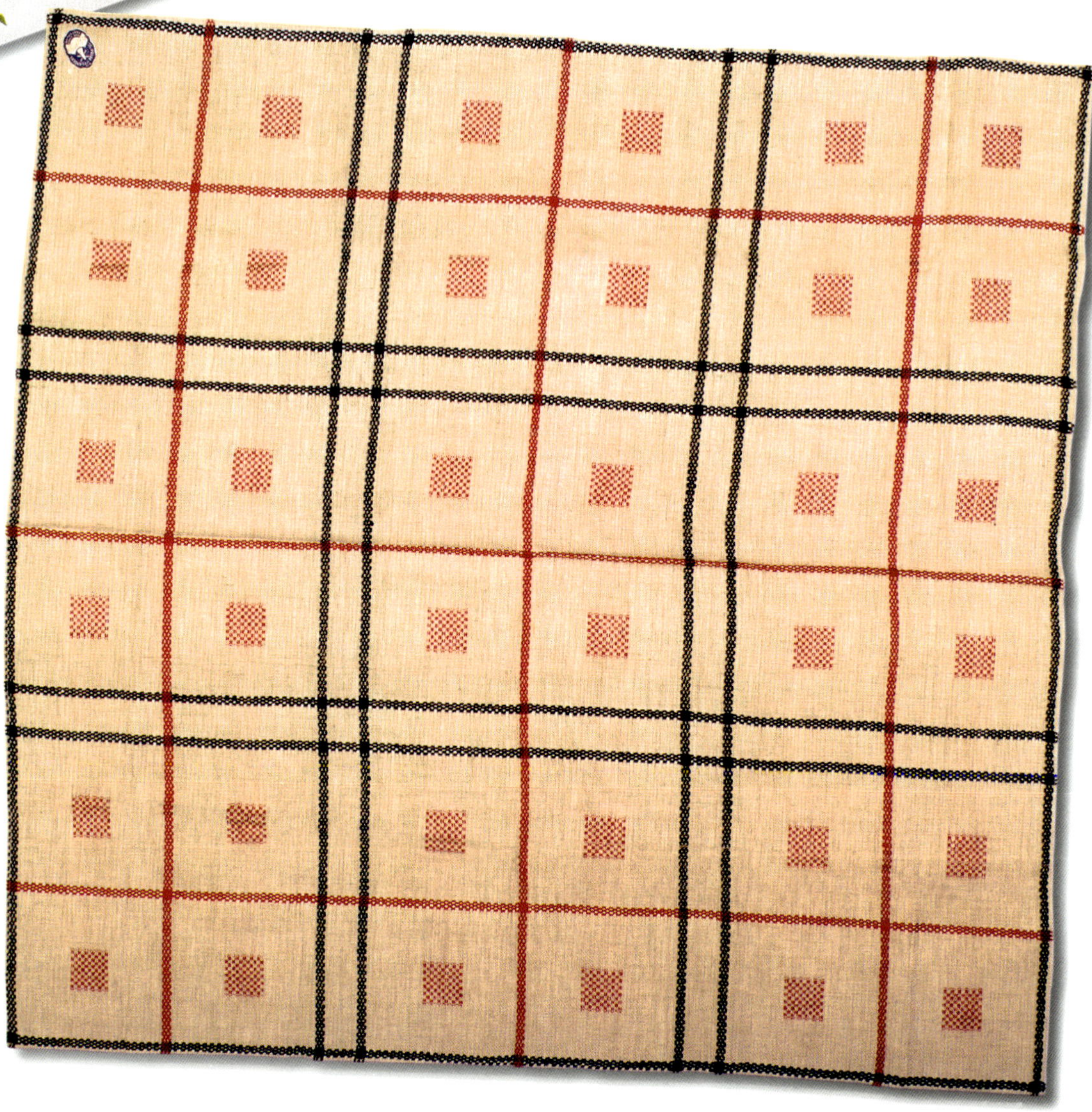

Pride of Flanders linen with six-inch squares covering the cloth and small checkerboard squares within. $18-25

Paper tag

Apple Blossom. Branches of pink and gray apple blossoms with green leaves displayed in groups of three across. $20-$35

Swoops of large poppies in corners, drifting out along sides. $20-$35

Purple pansies and golden yellow pansies with green leaves and gray shadows. This design was done in soft colors and has the appearance of a water color painting $35-50

Fairfield. Swags of flowers through the center with patches of posies along sides. $20-$35

Matching napkins. Set of four $20

Flowers and cherries on multi colored stripes. $20-35

Sunflower. Bright yellow sunflowers with burgundy centers and shaded green leaves set in squares - this glorious cloth is perfect for the summer table. $30-45

Napkins. Set of four $20

Additional color

The color combinations of yellow chrysanthemums with pink, red and yellow dianthus make this a real stand out. $30-45

Squares of soft green leaves with brightly colored tulips and other flowers. $20-$35

Take a look at this mod cloth in an all over pattern of bright oranges, yellow and lilacs. $18-25

Lush roses with yellow and blue daisies in three across squares, with dainty little sprigs around. $20-$35

Painted Daisy. Tossed clusters of daisies in shades of blue. $20-$35

Circles of large red poppies and tiny blue flowers on a yellow and white background. $20-35

All around border of large light and dark blue flowers with grass green leaves. $20-35

Clematis. Squares of blue clematis and spent clematis blossoms with climbing tendrils. $30-45

Close-up of **Clematis** in blue and in red. $30-45

Rosemead. Colorful squares showing flowers with cherries, strawberries and raspberries in corners. Artist: John Madsen. Design date: 1947. $20-$35

Rosemead napkins. Set of four $20

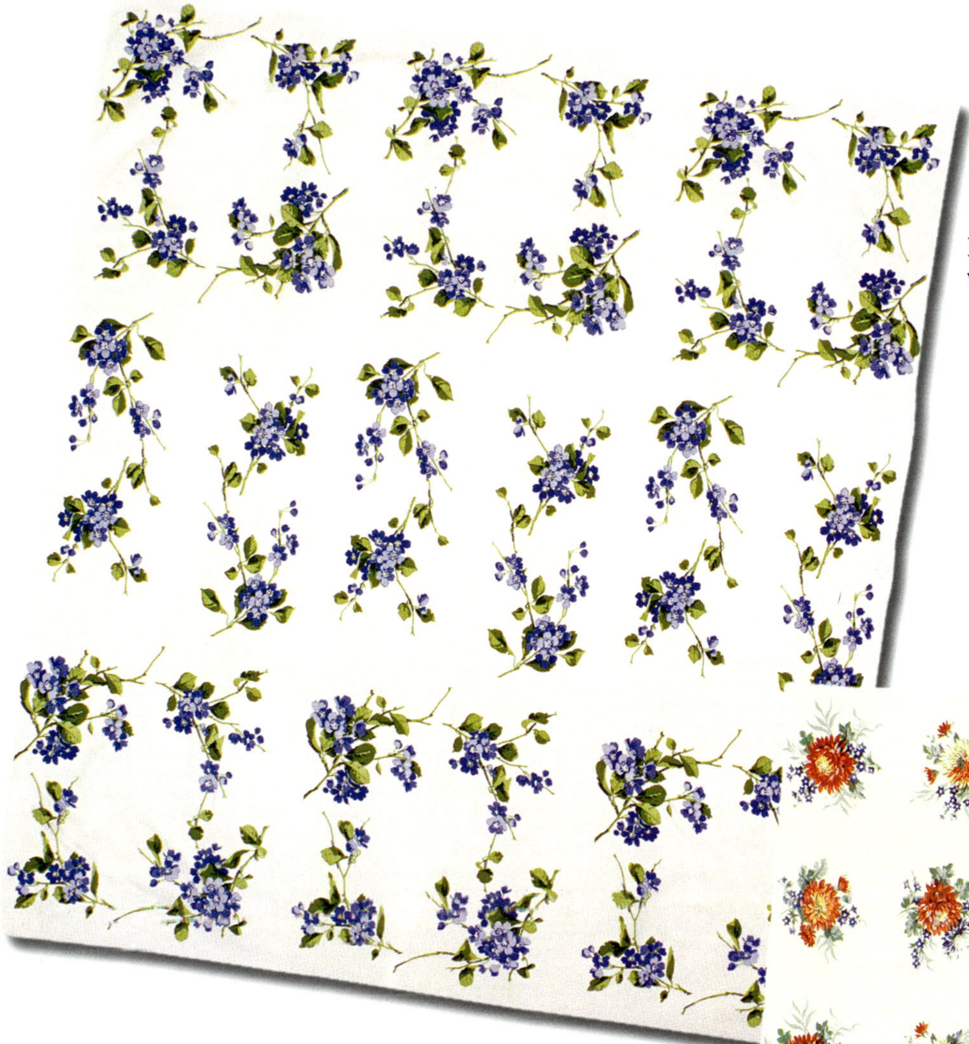

Branches of blue blossoms set in wreath like displays. $20-$35

Fluffy red chrysanthemums with dainty blue flowers and gray leaves. $20-35

Orange Blossom. White orange blossoms set within a white checked grid on a green background. $20-$35

Cosmos. Mixed floral clusters with red and gray daisies in a three across Design. $20-$35

Cosmos. Corner displays of primary colored flowers with soft green leaves and a striped border. Artist: John Madsen. Design date: 1940. $20-$35

Paper tag

Cosmos. Pink and lilac cosmos flowers with ferny green leaves. $20-35

Cosmos with a blue border and trailing flowers all around.

White flowers with red centers and green leaves on a gray background square. $20-$35

Dianthus. Dark green leaves with black shading and little puffs of blue flowers. $20-$35

Paper tag

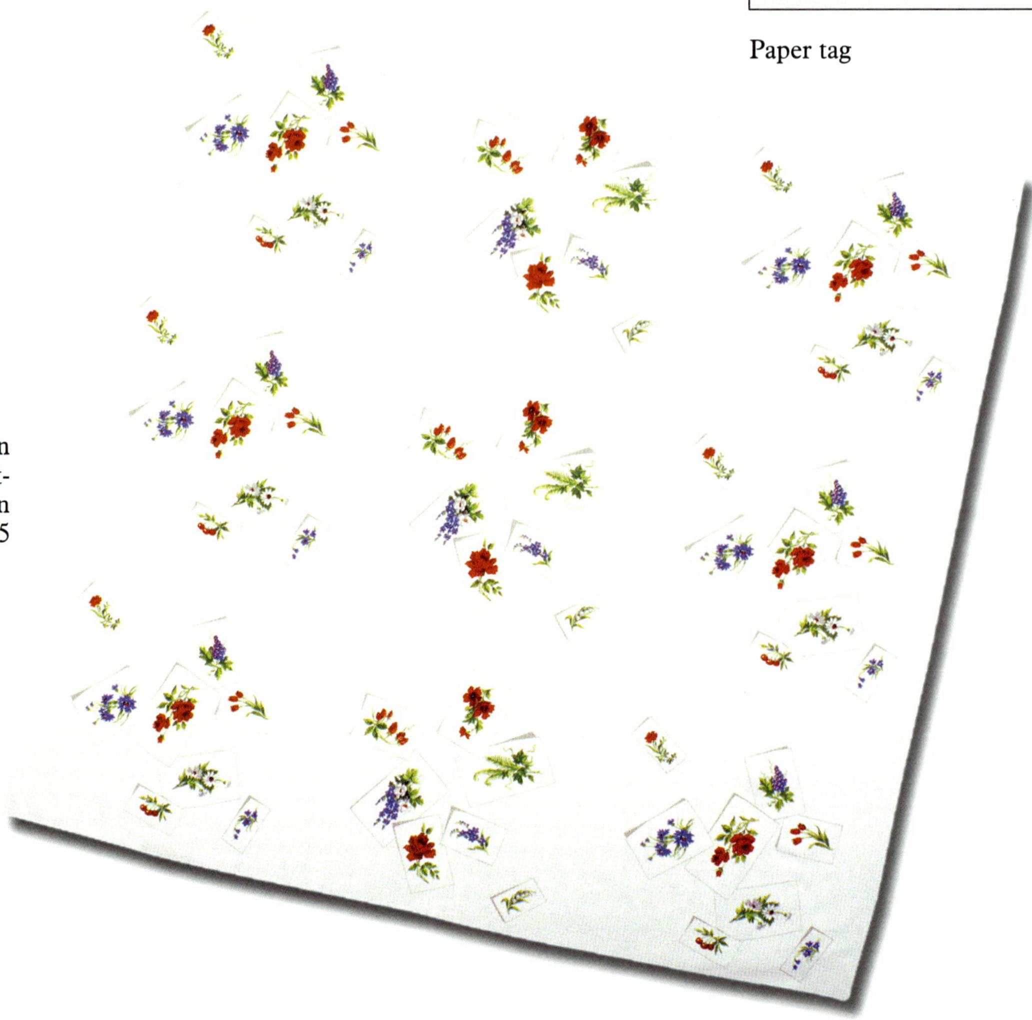

Delightful design of little "post-cards" of flowers in groups. $30-45

Paper tag

Flanders. Red poppies on a tan and white grid background. $20-$35

With flowers and cherries on a grid of yellow, green, blue, and red stripes, this cloth shows the brilliant colors for which Weil & Durrsè are famous. $20-$35

Detail of **Nasturtium** in blue and gray colors.

Nasturtium. Orange and yellow nasturtium flowers on a blue and white background. $35-50

Flora. Orange, yellow and brown flowers in an all over pattern. $20-35. *From the private collection of Pamela Glasell.*

Summerfield. Graceful floral displays highlighting red poppies and blue cornflowers. $20-$35

Napkins. Set of four $20

Pink chrysanthemums. $20-$35

Yellow flowers on twining brown branches. $20-$35

Tulip. Mauve colored tulip magnolias on white and yellow. $20-35

Tulip. A springtime scene of red tulips and yellow forsythia on a gray grid background. $30-45

Brightly colored flowers of red, blue, yellow and mauve on a green border. $20-$35

Meadowlands. Wipe clean tablecloth. Plastic with a polyester flannel backing. $20

American Beauty. Spectacular floral display, this beauty is a work of art. Weil & Durrsè also named one of their rose designs American Beauty, see page 80. $35-50

Paper tag

Red flowers with light and dark green leaves set in corner displays. $20-$35

Mixed floral clusters set in striped squares. $20-$35

Clusters of red, yellow and blue flowers with small blossoms in between. $20-$35

Pussy willows on pink background, brown branches and white catkins. $20-$35

Additional colors.

Burgundy background with gray stripes and white leaves. $20-$35

Paper tag

Blue Bell. Cheerful yellow bells mixed with tiny brown flowers. $20-$35

Flower pots with geraniums and tulips, coffee cups rooting ivy cuttings, flowers and fruit, all surrounded by "paint brush" swipes of color. $20-$35

Fourteen-inch squares of twining green leaves with richly colored red and yellow zinnias. $20-$35

Country Life. Gaily colored farmer and wife cloth in the Pennsylvania Dutch style. $20-35

Pink roses and blue larkspur. $20-$35

Yellow, blue and red flowers, each square bordered in blue. $20-$35

Stylized tulips in peach and brown. $20-35

Softly faded white pink gardenias on a green background. $20-$35

Garland. Border of colorful flowers with smaller flowers in the middle. $20-35

Red poppies, yellow achillea and gray and white daisies form 15-inch squares. $20-$35

Mixed flower bouquets alternating with green wheat. $20-35

Verbena. Red and white verbena with reaching stems of bright and dark green leaves. $30-45

Napkins. Set of four $20

Peach colored flowers with green leaves veined with white. $20-$35

Red poppies, blue cornflowers and bright yellow sheaves of wheat. $20-$35

Sweet little ladies slippers, Johnny jump ups, and more cover this cloth in floral garlands. Artist: Sergei Bogdanovich. Design date: 1957. $35-50

Phlox. Pink and red phlox, green shaded leaves, and little scattered blossoms. $20-$35

Phlox. Branches of alternating red and gray flowers in squares. $20-$35

Napkins. Set of four $20

Paper tag

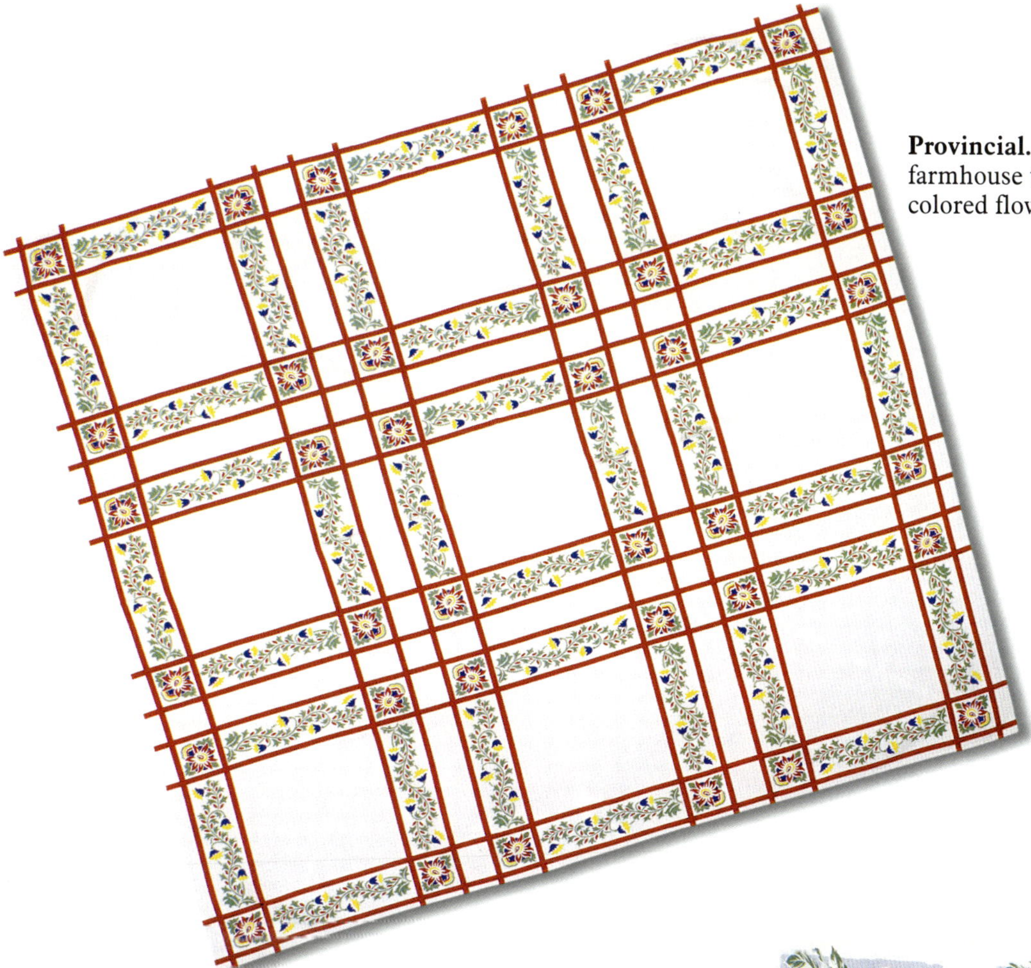

Provincial. So fitting on a French farmhouse table, this cloth has gaily colored flowers and vines. $20-$35

Paper tag

White daisies and pink peonies on a blue background. $20-$35

Red, yellow and blue flowers with green leaves form all around border and small squares in center. $20-35

This stunner shows red cineraria flowers with yellow shaded green leaves. Artist: Sergei Bogdanovich Design date: 1952. $30-45

Dahlia. Flowers in different shades of blue with leaves in the more modern olive green color. $20-$35

Paper tag

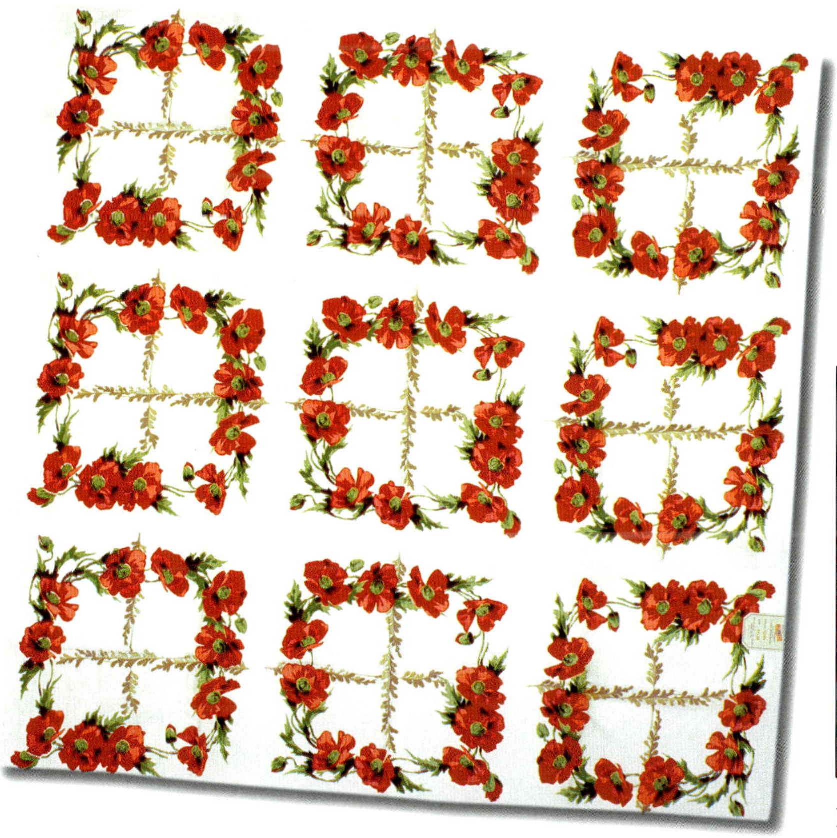

Poppy. Red poppies in squares with a gold leaf design in center. $20-$35. *From the private collection of Erin Henderson*

Note the color variation between these two paper tagged cloths. One shows leaves in light and dark gray while the leaves of the other cloth are tan and gold.

Poppy. Orchid colored poppies with green leaves and sprigs of babies breath. $20-$35

Paper tag

Paper tag

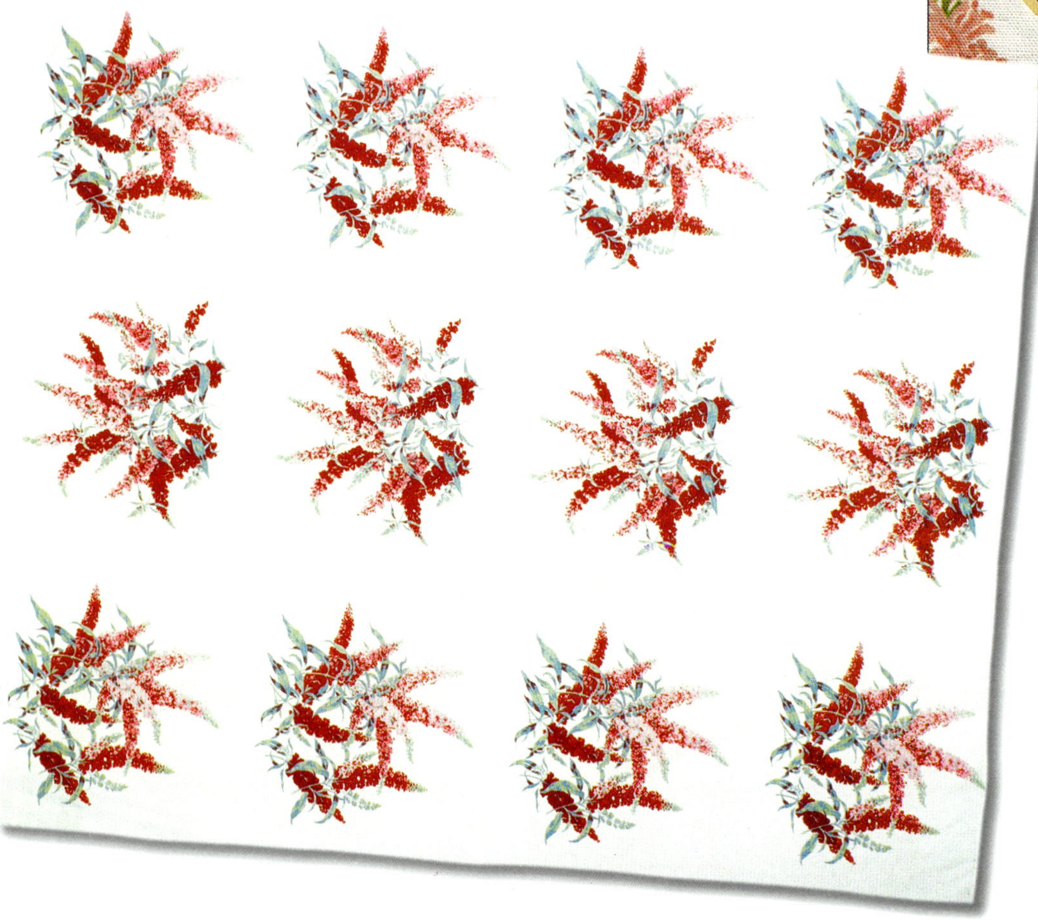

Butterfly Bush. Pink and burgundy flowers from the Buddleia plant. Artist: Sergei Bogdanovich. Design Date: **1957.** $30-45

Design in each corner of plums, flowers and leaves in burgundy and plum colors. $20-$35

Nosegay. Pansies, lily of the valley, bluebells and more decorate this cloth. $20-$35

Unusual design of cactus, palm trees, cannas and orchid mixed with others on an all around border. Artist: John Madsen. Design date: 1940. $30-45

Gingham Flower. Red gingham squares alternating with large red poppies. $20-$35. *From the private collection of Erin Henderson*

Paper tag

The ever-popular red poppy with blue bachelor buttons. $20-$35

Paper tag

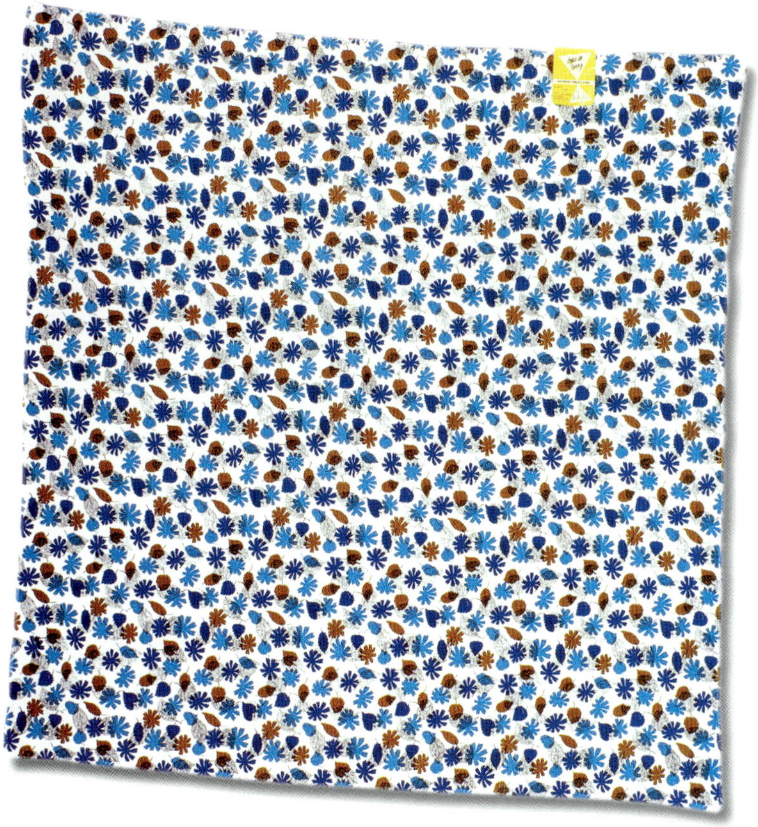

Leaf. Made of terrycloth, this has an all over pattern of blue, black and brown leaves. $18-25

Unusual combination of trillium flowers, white morning glory, and yellow and red toadflax . $30-45

Blueberry. Branches of red and blue berries set in 14 inch squares, with yellow border. $20-$35

Tahiti. Red background with white and gold leaf design. $20-$35

Paper tag

Red and yellow zinnias with airy lilac colored flowers and yellow green leaves. $20-$35

Red and gray leaves in a circle design repeated across cloth. $20-$35

Mixed floral border of Guernsey lily, linaria, dahlia, moon flower, and more looking as though they were painted in soft watercolors. $20-$35

Hibiscus. Red hibiscus with green leaves touched with white. $20-$35

A row of green leaves and red flowers on each side, with leaves sprinkled along the center. Gray and white striped background. $20-$35

Bouquets of flowers tied together with a yellow ribbon. Center row has squares of tossed single flowers. $20-$35

Russian design with peasants working in the field, peacocks, fairies, Cossacks and horses with carriage. $20-$35

Flower Festival. This is a polyester and cotton blend, permanent press cloth. Note the label is written in both English and French.

Paper tag

Dahlia. Blue dahlias. $20-$35

Magnificent display of larger than life pansies. Note the rich shading of the leaves. Artist: John Madsen. Design date: 1940. $35-50

Roses

Victoria. Stunning combination of black background, rich brown stems, and white roses. $20-35

Additional colors

Dutchess Rose. Pink and red roses with thorny brown branches and green shaded leaves. Artist: Sergei Bogdanovich. Design date: 1958. $20-$35

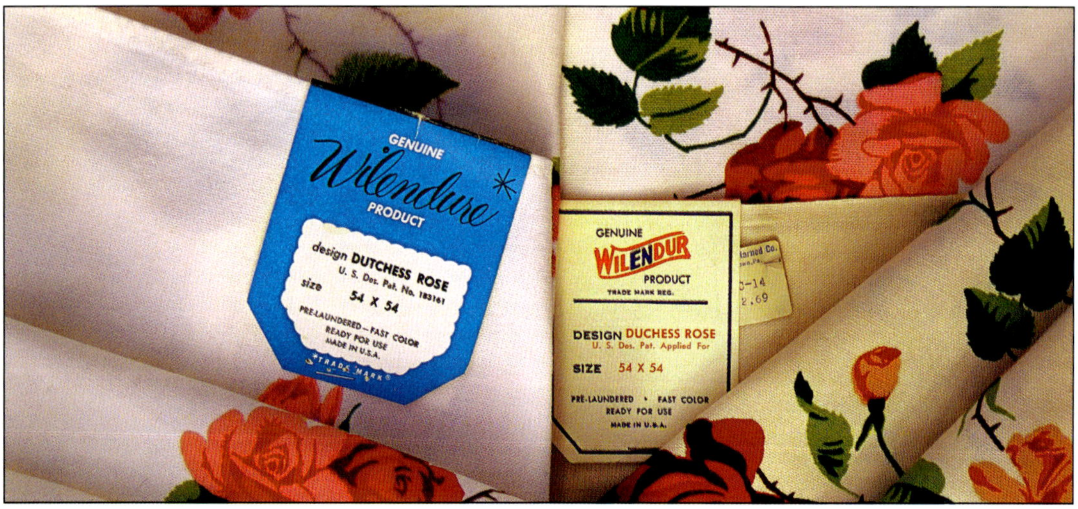

Paper tags from two **Dutchess Rose** tablecloths – note the different spelling of "Duchess". Also note the bright white background on one cloth with the other having a cream background. Wilendur cloth still has the original store price tag attached. The 54 x 54 inch tablecloth was priced at 2.89.

Empress Rose. Pride of Flanders pure linen bridge cloth with scalloped stitched hem. $20-$35

Empress Rose in yellow.

Paper tag

Empress Rose paper tag on terrycloth. This cloth retains the original price tag , from the "Garver Bros. Co." of Strasburg, Ohio, of $5.98.

Paper tag

Royal Rose. Large pink roses, green leaves and gray shadowing on a three across pattern. This cloth still has the original price tag pinned to it. Price for this 54 x 54 inch tablecloth was $3.95. Artist: Sergei Bogdanovich. Design date: 1953. $20-$35

Paper tag

American Beauty. Lush red American Beauty roses with green sheaves of wheat have made this one of the most popular of W & D Designs. $20-$35

Golden Dawn. Same design as the American Beauty rose with this one in yellow. $20-$35

Paper tag

Sharon. Yellow violets and wild roses set in squares surrounded by leaves, this is a shimmering beauty of spun rayon and cotton by America's Pride. This 54 x 54 inch cloth came as a set with 4 matching floral napkins, 17 inches square. $20-$35

Paper tag

Talisman Rose. Bright red roses and green leaves form a border around cloth. $20-$35

Paper tag

Floribunda. Small red roses with dainty rosebuds. $20-$35

Paper tag

Wildrose. Wild red roses with black shading and small green leaves set in 14 inch squares. $20-$35

Napkins. Set of four $20

Wildrose. Roses in blue and white with leaves the color of grass. $20-25

Paper tag

Shaggy Rose. Aqua colored roses with shaggy green leaves covers this cloth. This is made of a lightweight terry cloth. $20-$35

Paper tag

Shrub Rose. Red roses touched with feathery green ferns. $20-$35

Rambler. Red roses on a green and red grid background. $20-$35

Princess Rose. Sprays of pink Princess roses, this is a delicate beauty. This retains the original price tag from "The Fair" store in Chicago. The price on this 63 x 99 inch tablecloth was $7.70. Artist: Sergei Bogdanovich. Design date: 1956. $20-$35

Note the color variation between these two cloths. One has the typical green shaded leaves while the other has aqua blue leaves.

Additional colors

Tea Rose. This is a Setting Pretty design with yellow roses. $30

Paper tag

Lynbrook. Red roses scattered on a yellow grid background with smaller green and red lines. $20-$35

Long Stem Rose. A row of rosebuds on each side and a 16 inch wide column of red roses down the middle. Artist: Sergei Bogdanovich. Design Date: 1961. $30-45

This one is a beauty with large bright yellow roses, rosebuds, green leaves and accented by gray flowers in the background. $20-$35

Additional colors

Branches of wild roses along the sides with smaller flowers scattered down the middle. $20-$35

Margaret Rose. What a pretty one, perfect pink roses and green leaves. $20-$35

Paper tag

Violets and Geraniums

Alpine. Purple, lilac, and yellow violets mixed with white daisies and large green leaves. $30-$45

Bouquets of violets tied with a gay lilac ribbon along the sides, with tossed posies down the middle. $30-$45

Ivy Geranium. Red geranium flowers, green leaves burnished with a touch of brown, and white and yellow orchids with red stripes. $30-$45

Napkins. Set of four $20

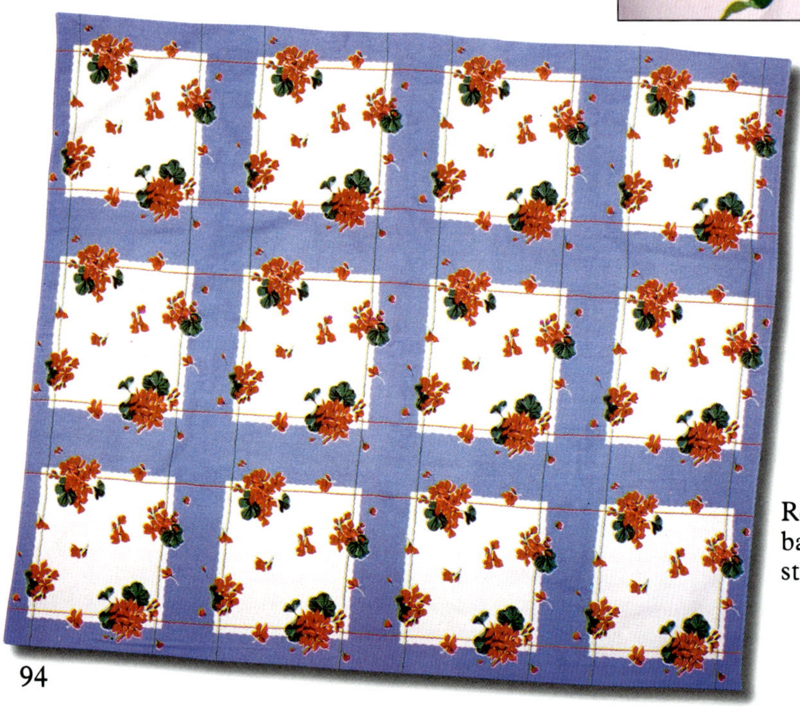

Red geraniums on white square background separated by blue stripes. $20-$35

Rich red geranium flowers, gray stems and two toned green leaves. $30-$45

Southwestern

Wonderful cloth showing life in a southwestern village. A senor with a donkey carrying a basket of fruit, mama with baby on her back, pottery bowls full of gourds, cactus and baskets of flowers, in six-inch squares. $20-$35

Additional colors

Mexicali. Colorful sombrero, blue pottery, striped scarf make the scene on this festive cloth. $20-$35

Manjares. Multi-color gourds and decorated pottery form the border of this cloth, edged with a rust-colored stripe. Artist: John Madsen. Design date: 1940. $20-$35

Mexican. Gourds, sombreros, and pottery with painted flowers on a red and white squares background. $20-35

Strokes of color frame strings of beads, stucco hacienda, Indian corn and shiny red pottery. $20-$35

Striped cloth with the look of an Indian blanket. $20-$35

Additional color

Mexican 2. Small squares showing seniors and senoritas on burros, oxen with cart, colorful pottery. Artist Name: Jacob Durrsè. Design date: 1937. $20-$35

Christmas and Fall

Snowy scene of street lights, holly berries, white poinsettias and acorns, all lit up by the North Star.
$30-$45

Paper tag

Merry Terry. Red poinsettia flowers, Christmas bells, a red ribbon and soft green leaves. $18-$25

Poinsettias, green shaded leaves and yellow berries on this 3-across cloth. $20-$35

Beautiful winter scene of evergreen trees with branches heavy with snow and snow capped mountains in the background. $35-$50

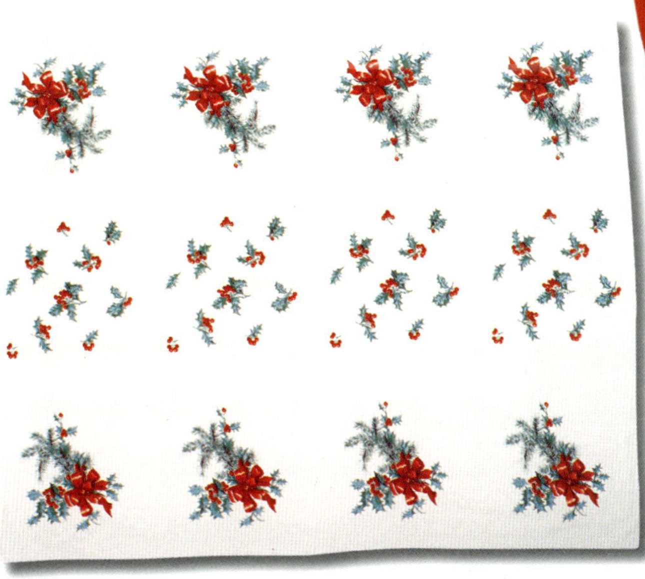

Holly, spruce branches and a big red bow decorate groups on sides of cloth with little mistletoe berries down the center. $20-$35

Noel. An inviting home scene with smoke coming out of the chimney, lights in the windows and a displays of white branches, acorns and shiny gold Christmas tree ornaments. $35-$50

A popular Christmas plant, the poinsettia is shown here on a white background with gray and white stripes. $20-$35

Pine branches and roses in white and gray on a red background. $30-$45

Solid red sides with a white band down the middle showing poinsettias and holly. $20-$35

Poinsettia Imperiale. This 1960s tablecloth is a light weight blend of 50% polyester and 50% cotton. These cloths were packaged in plastic and included a paper label which gave the design name and washing instructions. $18-$25

Corner design of spruce boughs, tree ornaments and red bow. $20-$35

Christmas scene with a touch of the Dutch look to it set in 12 inch circles with a floral design along the sides. $20-$35

Harvest season cloth with brightly colored candy corn, sweet peas plump in their pod and striped gourds. $35-$50

Yellow sunflowers, orange bitter-sweet, with grapes, cherries and wheat, perfect for fall. $20-$35

Paper tag

Autumn. Brown and green ivy with tendrils and yellow leaves. Note this matches the "Ivy" design but in browns instead of greens. Artist: Sergei Bogdanovich. Design date: 1955. $20-$35

Buckeye. Branches of green leaves and brown chestnuts in their fuzzy pods. $30-$45

Dogwoods and Morning Glories

Additional colors

Dogwood. Stark black background with white dogwood flowers touched with brown and gray. Artist: Sergei Bogdanovich. Design date: 1952. $20-$35

Paper tag

Lakewood. Yellow dogwood flowers in a 3-across design. $20-$35

Morning Glory. Orange morning glories on a bright yellow and white grid background. $20-$35

Pink morning glories and grass green leaves edge around sides of cloth. $20-$35

Additional colors of red morning glory cloth that is shown in "Pattern Examples". $20-$35

Red and pink morning glories and yellow buds on a gray and white grid. $20-$35

Light and dark green leaves with red and pink morning glories. $20-$35

Fruits and Vegetables

Cherry. Clusters of rich red cherries with red streaked green leaves. $30-$45

Cherry. Two rows of shiny red cherries and blossoms on each side of cloth with a center design of blossoms with little sprigs of cherries. Artist: John Madsen. Design date: 1947. $30-$45

Red cherries still on the branch alternates with clusters of cherry blossoms, all in 7 inch squares over cloth. $30-$45

Crabapple in red color.

Crabapple. Yellow crabapples and moss green leaves border on this terry tablecloth. $18-$25

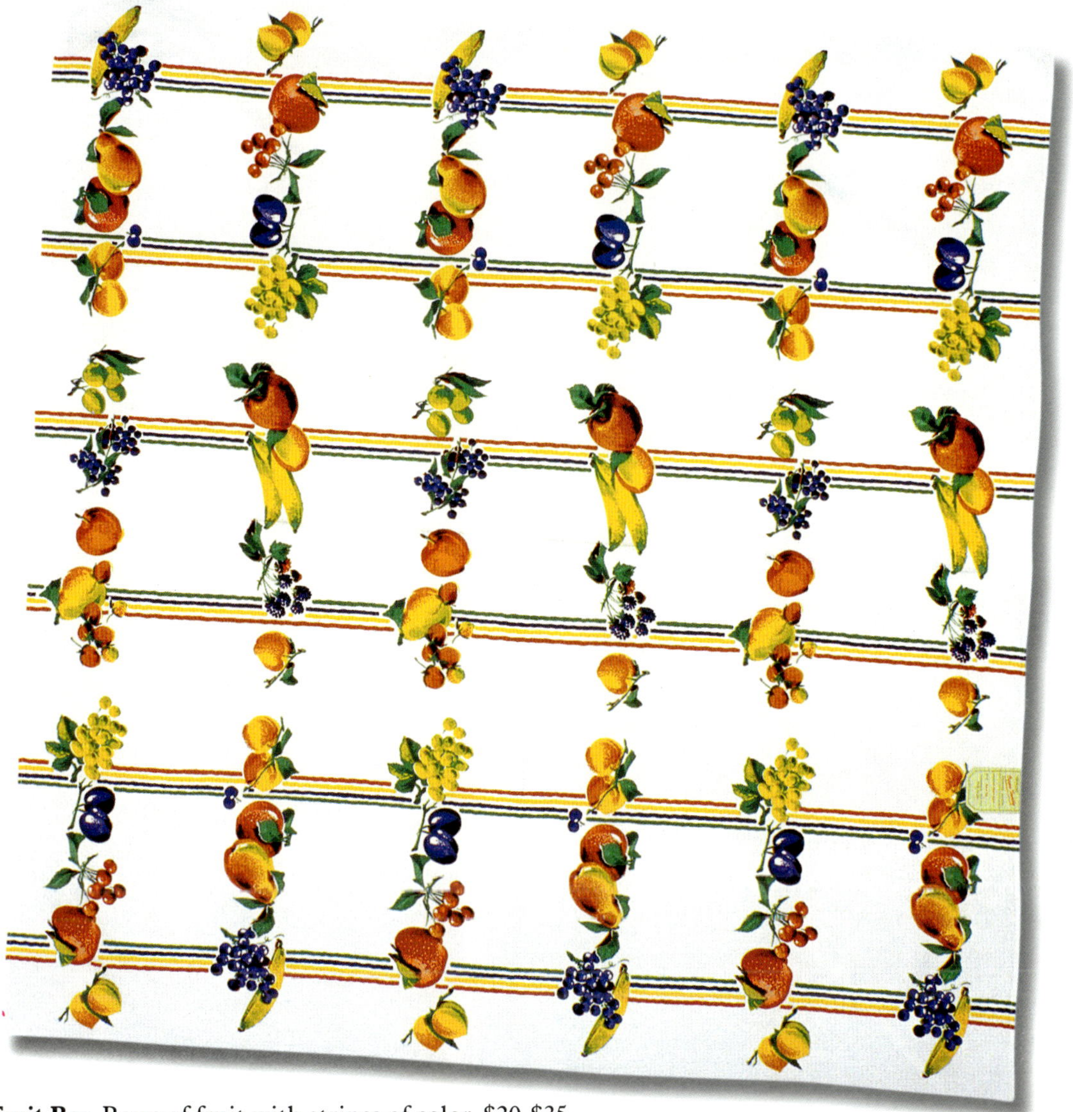

Fruit Bar. Rows of fruit with stripes of color. $20-$35

Paper tag

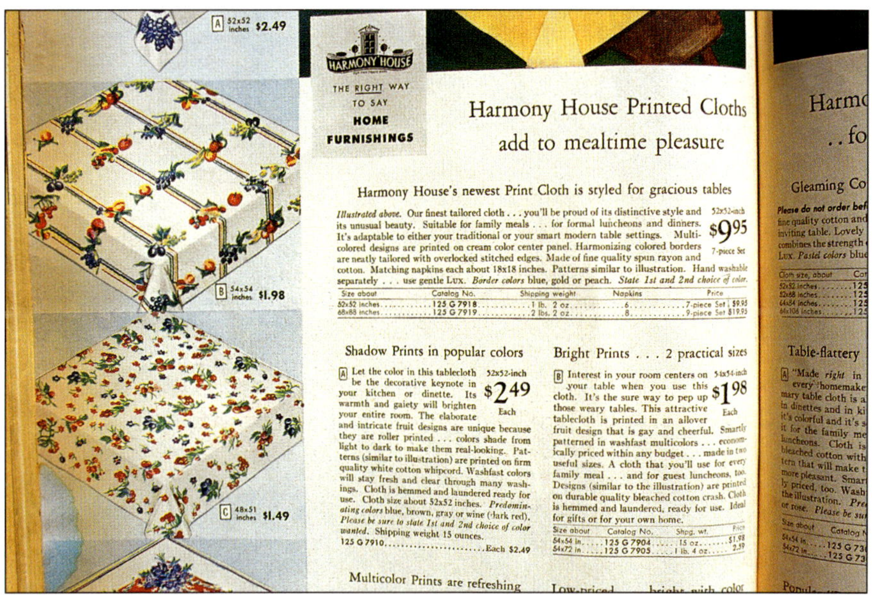

THE RIGHT WAY TO SAY HOME FURNISHINGS

Harmony House Printed Cloths add to mealtime pleasure

Harmony House's newest Print Cloth is styled for gracious tables

Illustrated above. Our finest tailored cloth . . . you'll be proud of its distinctive style and its unusual beauty. Suitable for family meals . . . for formal luncheons and dinners. It's adaptable to either your traditional or your smart modern table settings. Multi-colored designs are printed on cream color center panel. Harmonizing colored borders are neatly tailored with overlocked stitched edges. Made of fine quality spun rayon and cotton. Matching napkins each about 18x18 inches. Patterns similar to illustration. Hand washable separately . . . use gentle Lux. *Border colors* blue, gold or peach. *State 1st and 2nd choice of color.* 52x52-inch $9.95 7-piece Set

Size about	Catalog No.	Shipping weight	Napkins	Price
52x52 inches	125 G 7918	1 lb. 2 oz.	6	7-piece Set $9.95
68x88 inches	125 G 7919	2 lbs. 2 oz.	8	9-piece Set $19.95

Shadow Prints in popular colors

A Let the color in this tablecloth be the decorative keynote in your kitchen or dinette. Its warmth and gaiety will brighten your entire room. The elaborate and intricate fruit designs are unique because they are roller printed . . . colors shade from light to dark to make them real-looking. Patterns (similar to illustration) are printed on firm quality white cotton whipcord. Washfast colors will stay fresh and clear through many washings. Cloth is hemmed and laundered ready for use. Cloth size about 52x52 inches. *Predominating colors* blue, brown, gray or wine (dark red). *Please be sure to state 1st and 2nd choice of color wanted.* Shipping weight 15 ounces. 52x52-inch $2.49 Each

125 G 7910 . . . . . Each $2.49

Bright Prints . . . 2 practical sizes

B Interest in your room centers on your table when you use this cloth. It's the sure way to pep up those weary tables. This attractive tablecloth is printed in an allover fruit design that is gay and cheerful. Smartly patterned in washfast multicolors . . . economically priced within any budget . . . made in two useful sizes. A cloth that you'll use for every family meal . . . and for guest luncheons, too. Designs (similar to the illustration) are printed on durable quality bleached cotton crash. Cloth is hemmed and laundered, ready for use. Ideal for gifts or for your own home. 54x54-inch $1.98 Each

Size about	Catalog No.	Shpg. wt.	Price
54x54 in.	125 G 7904	15 oz.	$1.98
54x72 in.	125 G 7905	1 lb. 4 oz.	2.59

Multicolor Prints are refreshing

1946 Sears catalog showing tablecloths. The catalog reads, "This attractive tablecloth is printed in an all over fruit design that is gay and cheerful." This 54 x 54 inch tablecloth was priced at $1.98.

Two rows of red and yellow apples down sides with center pattern emphasizing the yellow apple blossoms. Note the similarity of this pattern to the "Cherry" cloth. $25-$40

Berry. Blue background accented with red and darker blue with displays of mixed berries on a 12 inch white square. $25-$40

Mixed fruits in a three across design. $20-$35

Big shiny apples, blue and yellow grapes, pears and cherries make up this still life of fruit. $25-$40

Plum. Blue plums, black berries and green leaves. $20-$35

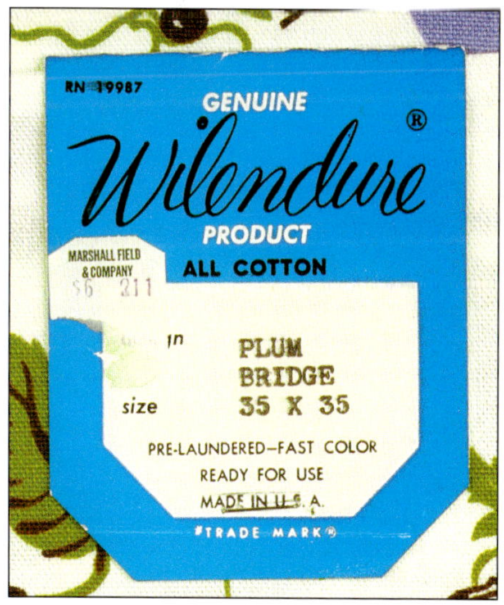

Paper tag

Additional color

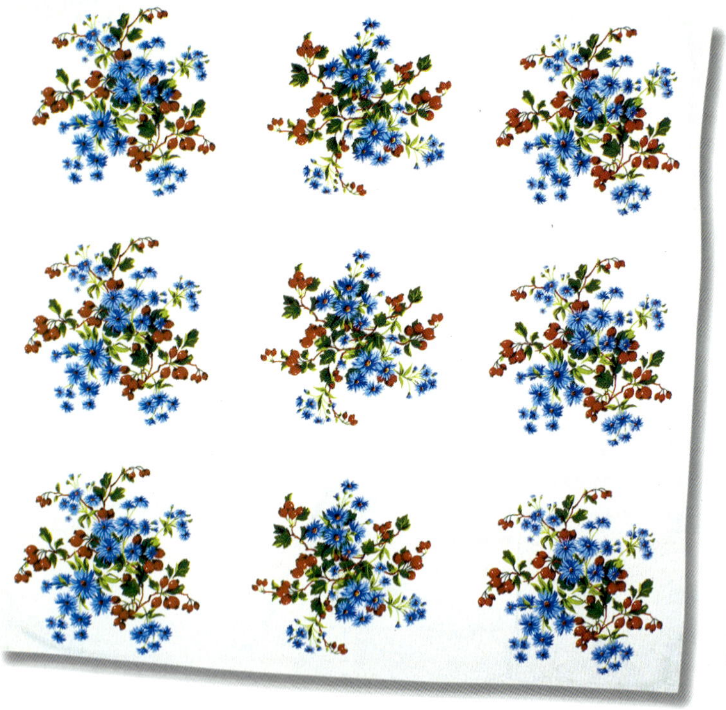

Spectacular cloth showing blue daisies and red berries on a bright white background. $25-$40.

Melon, strawberries, bananas, and more in large, colorful groups.

Fireberry. Cream and tan grid background with 5 inch squares of tiny red fireberries and white and red blossoms. This cloth is a good example of the type of damage that occurs to fabric that has been stored improperly. Note the light brown area around the paper tag. This shows that the cloth was folded and stored with just this area exposed to the air. Dust accumulated on this for so many years that even with proper de-staining and laundering techniques used, the darker color remains. The dust and insect damage has compromised the integrity of the fabric itself and left it with small holes. $20-$35

Watermelons, apples, and cherries with bright yellow border.

Red raspberries and blue raspberries. This is a 44-inch wide by 87-inch long piece of fabric that was cut off of the bolt. The sides are selvage and the ends uneven and ready to be hemmed. $40

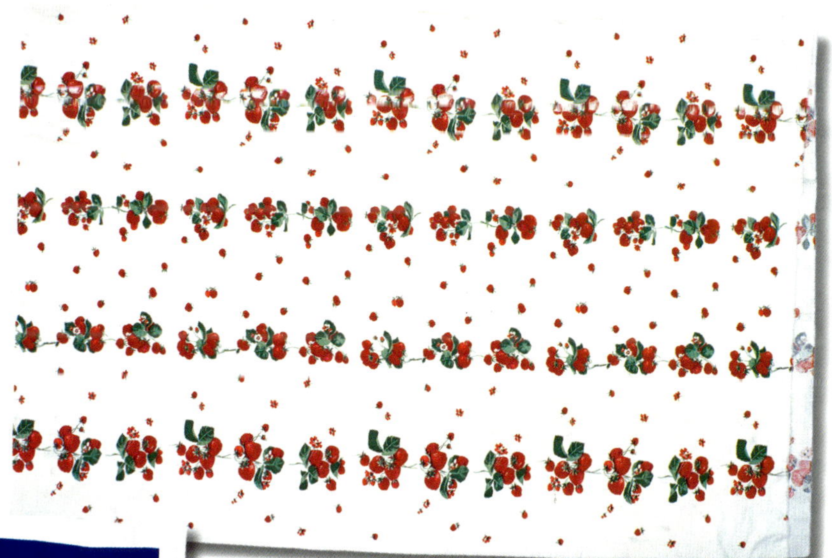

This fabric was cut from the bolt and measures 54 inches wide by 14 feet long. These large pieces were often used to create living room curtains or cut up and sewn into a tablecloth with matching kitchen curtains. $70

The ultimate fruit tablecloth this one has gooseberries, apricots, grapes, bananas, cherries, plums and pears in rich colors, all set in 6 x 7 inch squares. $20-$35

Note the similarities in the previous cloth to this one with both having small multi-colored squares of fruit. This one has an all-over design.. $20-$35

Corner pattern of a bowl of plums with pineapple and mixed fruit all around, an outer border of berries, flowers, and leaves and little fruits scattered in the center. $20-$35

Blueberries, cherries, and blue flowers in small, yellow-bordered squares. $20-$35

Green grapes and leaves with twining tendrils. $20-$35

Red raspberries with gray leaves. $20-$35

Branches of colorful red and yellow apples.

Barberry. Red berries with fluffy white flowers alternate with dainty yellow bell flowers. $30-$45

Strawberry. Fresh strawberries with white blossoms and green leaves. $30-$45

Large displays of yellow apples and orange berries. $20-$35

Border all around of red strawberries with yellow blossoms and strawberries scattered in the center. $20-$35

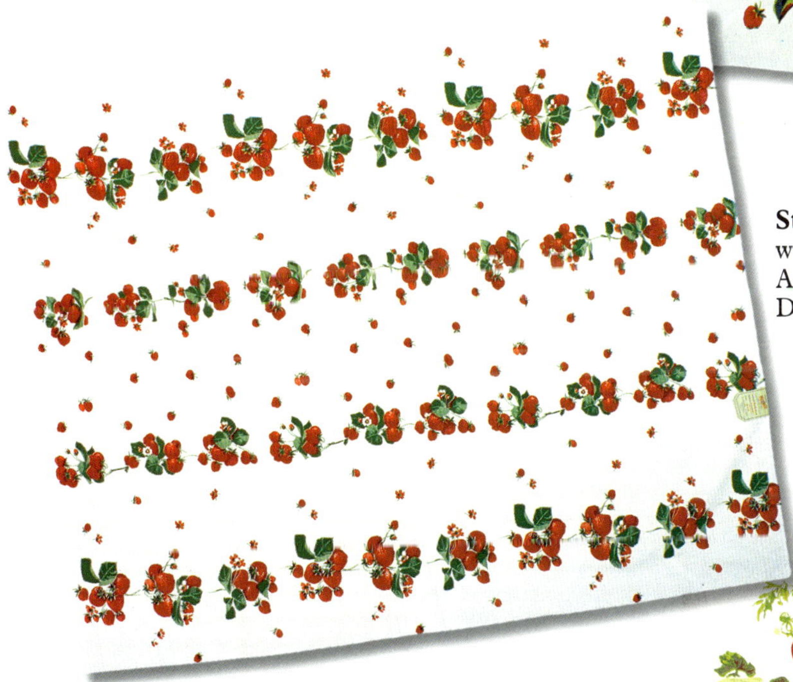

Strawberry. Rows of strawberries with red and white blossoms. Artist: John S. Madsen. Date of Design: 1944. $30-$45

Kitchen Garden. Vines of ripe tomatoes and green gourds attached to a wooden post with a red string. $30-$45

Vegetable cloth with cabbage, plump orange carrots, eggplants, and more in an all-around border. Artist: John Madsen. Design date: 1939.

Beautiful cloth with a tropical flavor showing mangos, pineapples, anthurium, hibiscus, and orchids. $20-$35

Unusual

This tablecloth was used by Weil & Durrsè to illustrate their Pride of Flanders product line. This measures 42 x 62 and shows the Weil & Durrsè insignia at the top and "Pride of Flanders" written across the cloth. Along the bottom it reads "Pure Linen, Originality In Prints, Fast Cloth". A beautiful cloth which advertises Pride of Flanders linens and shows the poppy fields of Flanders. Weil & Durrsè obtained their linen from Flanders, Belgium. $150

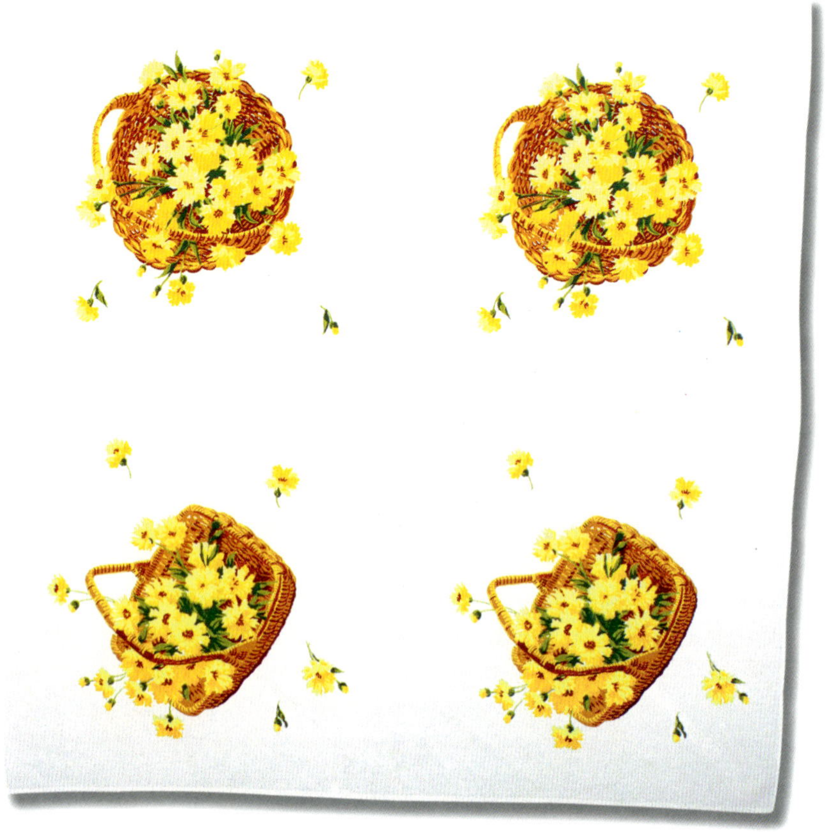

Wicker baskets, round and square, filled with yellow dianthus. $20-$35

Ceramic. A modern 1950s cloth with a southwestern flair. Pottery in earth colors. $30-$45

Paper tag

Wild Ducks. Ducks flying and swimming, reeds of grass, cattails and water lilies. $30-$45

Paper tag

Stripes of dark brown, orange, light brown, and yellow. $20-$35

Black iron trivets, with pink, gray and red pottery bowls and spoon. $20-$35

Additional color.

Chicken. Colorful roosters and hens with baby chicks. Note the corners showing mama chicken with her baby chicks around her and a little one perched on her head. $45-$60

Paper tag. This tablecloth has an America's Pride paper tag attached and also has the Wilendur sewn-in cloth label. This was a factory error. Earlier America's Pride tags were printed with asterisks, as such: "America's Pride". The later ones dropped the asterisks, as shown in this tag. Most America's Pride cloths were a cotton/rayon blend, but as shown here they also produced a pure cotton cloth.

Additional color

Italian themed cloth with grapes, jugs of wine, guitar, rooster, and a banner with the colors of the Italian flag. $30-$45

Wooden salad bowls, jars of vinegar and oil, olives and lemons with a sprig of dill. The owner of this cloth crocheted a multi color border of tatting to match the colors of the design. This beautiful hand made addition is evidence of care and appreciation of the cloth and makes it all the more special. $20-$35

Decorative fruit bowl, salt and pepper shakers, cheese and crackers, coffee cup and spoon. $20-$35

Liquor Labels. What a fun cloth with pictures of liquor labels, including such names as Black Head Rum, Old Grand Dad, Gilbeys Gin, and Champagne Montebello. $35-$50

Paper tag

Paper tag

Cat-tail. A favorite cloth of cattail lovers, this shows spikes of cattails, beautifully shaded green leaves, and fronds of golden wheat. $35-$50

Wonderfully different cloth, this design has an all over mosaic tile appearance, including the richly colored red, yellow, and orange zinnias. $20-$35

Paper tag

Penn-Dutch. Folk art with a Germanic influence, this cloth's design influence is the artwork of the Pennsylvania German people. A popular design showing hearts, which symbolize love, and birds, which bring good luck. Artist: Sergei Bogdanovich. Design Date: 1957. $20-$35

One of the most popular Weil & Durse designs showing bright red lobsters and gray shelled clams, with aqua green seaweed in the background. Artist: Sergei Bogdanovich. Design Date: 1958. $40-$65

An unusual design showing different mementos, such as a small silhouette portrait of a gentleman, a compass, a little perfume bottle, a worn picture of a pretty woman, and two tickets to the ballet. $35-$50

Look-alikes

For completeness, a few look-alikes are included, which are products that are known from research to be from other companies. The intent is to help the reader avoid mistakes of identification.

Weil and Durrsè linens had no competitor to match their beauty, but there were other textile companies whose tablecloths had many similarities. Here are a few of the tablecloths that are commonly mistaken for Wilendur.

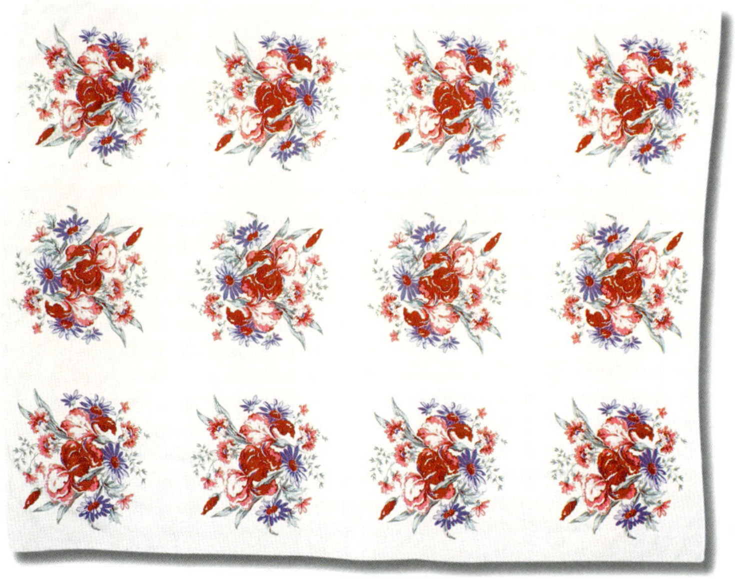

St. Regis, a Springmaid product. Similarities are the mixed floral bouquets and three-across pattern.

Sun-Glo. Similarities are the richly colored roses, olive green stems typical of later Wilendures, sturdy cotton cloth, and large squares of pattern repeating across the cloth.

California Hand Prints. Similarities are the three-across pattern, squares bordered in ivy, and rich colors. California Hand Prints tablecloths were often made of the same sail cloth as Weil & Durrse's tablecloths.

Thomaston Pedigree. Similarities are the large squares of florals across, bright colors, smaller flowers scattered around, and three-across pattern.

Tablecloth Sets

The Weil & Durrsè Company produced various tablecloth sets with coordinating napkins. These were often packaged in attractive boxes, with many different designs to the boxes themselves. The tablecloths were in various sizes, all with napkins included. Some sets would have a solid color tablecloth with floral napkins, while others had a floral tablecloth with solid napkins. The sets were named according to the three table sizes they were intended for:

"Festive," 45" x 45"
"Bridge," 34" x 34"
"Festive Bridge," 36" x 36"

They would also name the same-sized sets "Festive Tea Set" and "Tea Time."

Yellow Princess Rose with matching floral napkins. Note the yellow hem stitching on both the cloth and the napkins. $20-$35

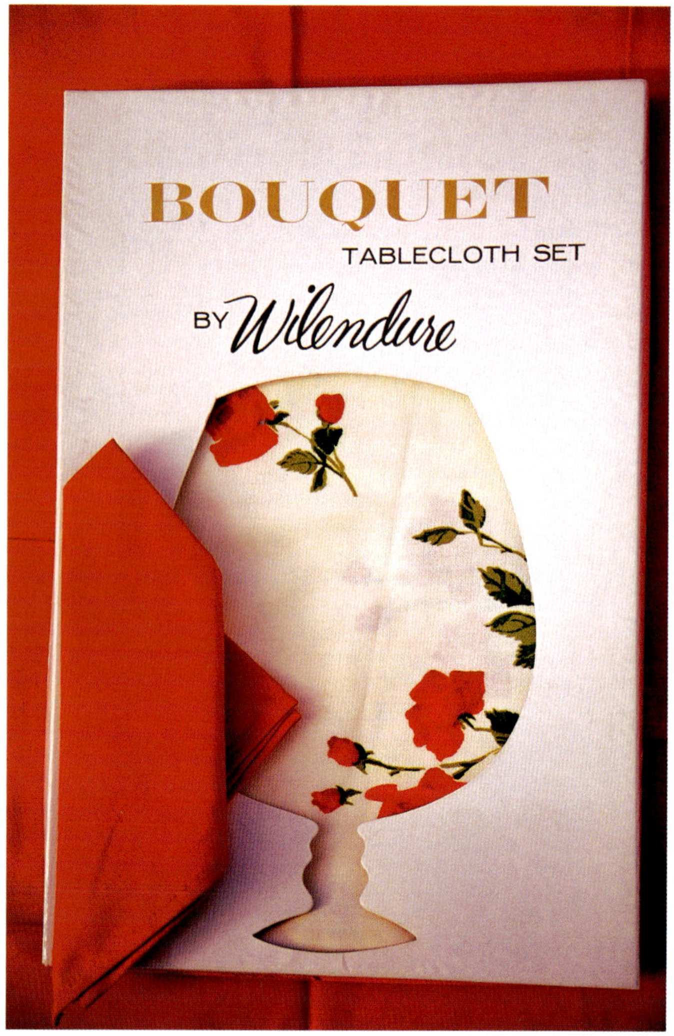

Floribunda. Called a Bouquet set, this came with four solid red napkins. Note the cut out design of the box. Boxed set $50-70

Tea Time set. Modern design of orange florals. Box has cut-out windows in the shape of a tea cup and tea pot. Boxed set $35-$50

Granada. "Luxurious Tablecloth Set by Wilendure" printed on front of the box. Note the cut-out design of the box styled to look like a plate with the silverware on the sides. Boxed set $35-$50

Tea Time. Size 45 x 45 inches. Solid orange tablecloth with four floral napkins. Napkin pattern shows orange morning glories climbing a wooden post. Paper tagged $20-45

Pom Pom. Boxed set of Pom Pom design tablecloth in red floral with four solid red napkins. This set is priced at the higher end of the price range because of the beautiful floral design, bright colors, and the original box is included which has an imprinted floral design on the front. $75-$100

Princess Rose tablecloth with six matching napkins. No need to buy wrapping paper for this gift as Weil and Durrsè presented this set so beautifully with the layered napkins and shiny white ribbons. So perfect, straight from the department store shelf to the birthday table. Paper-tagged set $80-$100.

Flora. Box set of tablecloth with four napkins. Note cut out design of box. $40-$60.
From the private collection of Pamela Glasell

Windblown. Mixed flowers in blues with windblown petals. Packaged attractively with napkins folded into the tablecloth and finished off with two white ribbons. Paper tagged $30-$50

Runners, Placemats, Napkins, and Curtains

Weil and Durrsè produced placemats, runners, napkins and curtains in designs and colors to match their tablecloths. The table linens were sold individually allowing the housewife to purchase the exact number needed for her family. Napkins came in sizes of 16 x 16 inches and 12 x 14 inches. Placemats were 11 x 16 inches. Weil and Durrsè sold boxed sets of 4 placemats with 4 matching napkins that were called Luncheon sets. Runners were sold in several sizes including 13 x 70, 13 x 90, 11 x 34, and 17 x 54 inches. Kitchen towels were sized 16 x 28 and 16 x 32 inches. The smaller tea towels were sized 13 x 18 and 9 x 14 inches.

Strawberry. This tablecloth is made from Duracloth, which is a medium-weight cotton with a very smooth finish. Boxed set $80-$100.

Placemats in the **Meadowbloom** design. These placemats measure 11 x 16 inches and each has the original price tag attached. They were priced at .29 cents each. Set of six $25

Same design in mixed colors

Runner, placemats, and napkins purchased separately to go with matching **Meadowbloom** tablecloth in peach.

Mixed napkins

Mixed napkins in cheerful reds and yellows

Mixed napkins with fall leaves, magnolias, and classical white.

Set of twelve **Princess Rose** napkins, all with original price tags from "The Fair" store in Chicago. Originally priced at 39 cents each. Set of 12 $22.

Weil & Durrsè sold these curtains to match the **Petunia** tablecloth. Sold as one piece, this has a ruffled green and white valance and trim. $20.

Runner in the **Holly and Pine** design. Runner measures 17 x 54 inches. $25

Chapter 2
Towels

Branches of flowers and berries with red borders. $15

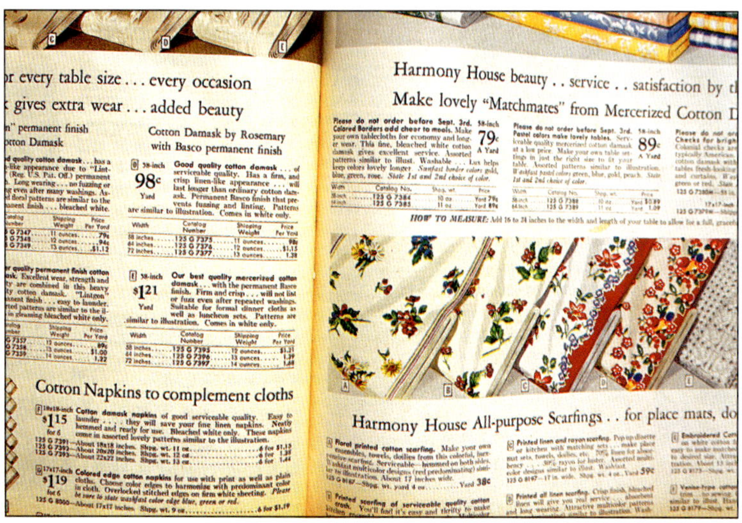

1946 Sears catalog showing toweling fabric. The catalog reads, "Make your own ensembles, towels, doilies from this colorful, inexpensive scarfing." Priced at 38 cents a yard.

Birds and butterflies on cherry blossom branchcs. $15

Nasturtiums in red and blue with green border. $15

Linen in a vegetable design with yellow string beans down the sides. Artist: John Madsen. Design date: 1939. $15

Mixed floral display with posies scattered all around the border. $15

Pride of Flanders linen showing a collection of old time blue glass lamps. $15

The creative talents of the Weil & Durrsè artists is demonstrated in this window design of ruffly curtain, red wooden shutters, and flower pots of cyclamen and begonias on the window sill. $20

Dainty linen tea towel with burgundy and white gardenias. $10

Gardenia. White gardenias on a lime green background. This is often mistaken as a rose design. $15

Fruit, water jug, and flowers all in pink. $15

Window design with branches of apples and apple blossoms. Yellow flowering plant on the window sill, ruffled valance across the top and red brick trim all around. $20

Red, blue and yellow morning glories on bright yellow border. $15

Apples and grapes on both ends with sprigs of fruit along the sides. Artist: John Madsen. Design date: 1940. $15

Antiques towel in linen. $10

Fluffy white flowers and hibiscus. $15

Glorious grapes and yellow shaded leaves. Beautiful hand crocheted trim finishes the ends. $15

Fruits and florals in shades of purple and plum. $10

Red poppies in a bowl with yellow terra cotta pitcher decorated with flowers. $15

Girl churning butter in an old time kitchen scene. $10

Cactus growing in a cute burro planter, white and blue daisies in a dish, and a border of red gingham. $20

Blue and white daisies on a pink background. $10

Flowers growing in a red pot and a pig planter
with ivy cutting trailing from it. $20

Flower pot with pink geranium, book, and little piglet salt and pepper shakers.
Note the same red floral design this artist gives to the animal designs. $20

Colorful red, blue and yellow floral display with a gray shadow border. $15

Yellow flowers with blue larkspur. $15

Fireberry design with red berries on blue and white background. Often mistaken for cherries.

Towels with a Russian influence. Girl wearing colorful scarf and apron carrying water buckets. Well and cabins in background. $15

Young man and woman performing a traditional Russian folk dance, with a Russian castle in the background. Colorful ribbons stream from the girl's hair and she holds a yellow hankie in her hand. $15

Golden samovar with a tiny teapot on top, and surrounded by colorful pottery items. $15

Spanish Tile. Squares of fruit in earth tone colors. $10

Red and yellow apples on green branches with bright stripes between. $15

Hibiscus and poppies divided by green and red stripes. $15

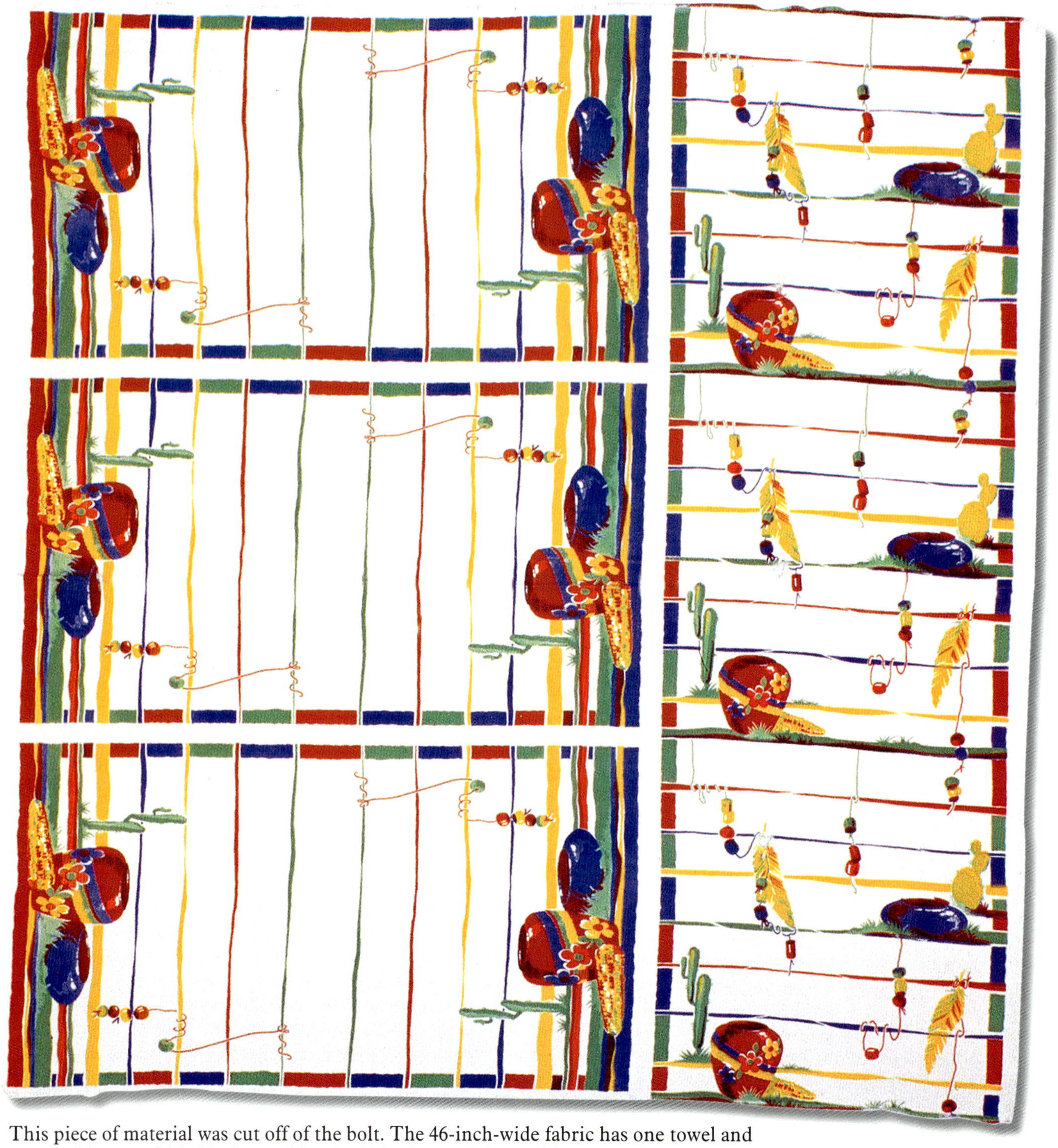

This piece of material was cut off of the bolt. The 46-inch-wide fabric has one towel and one napkin per width. Each towel is 16 x 28 inches and the napkins are 16 inches square. The customer could purchase whatever length she needed, depending upon how many towels and napkins were desired. Cutting and hemming was all that was needed to provide colorful, ready- to-use kitchen linens. $60

Pottery and gourds in cheerful fiesta colors. $15

Green, orange and yellow flowers with brown leaves. $15

Small figures of woman on both ends, all in black and white. $15

Blue lilacs with twining green leaves. $15

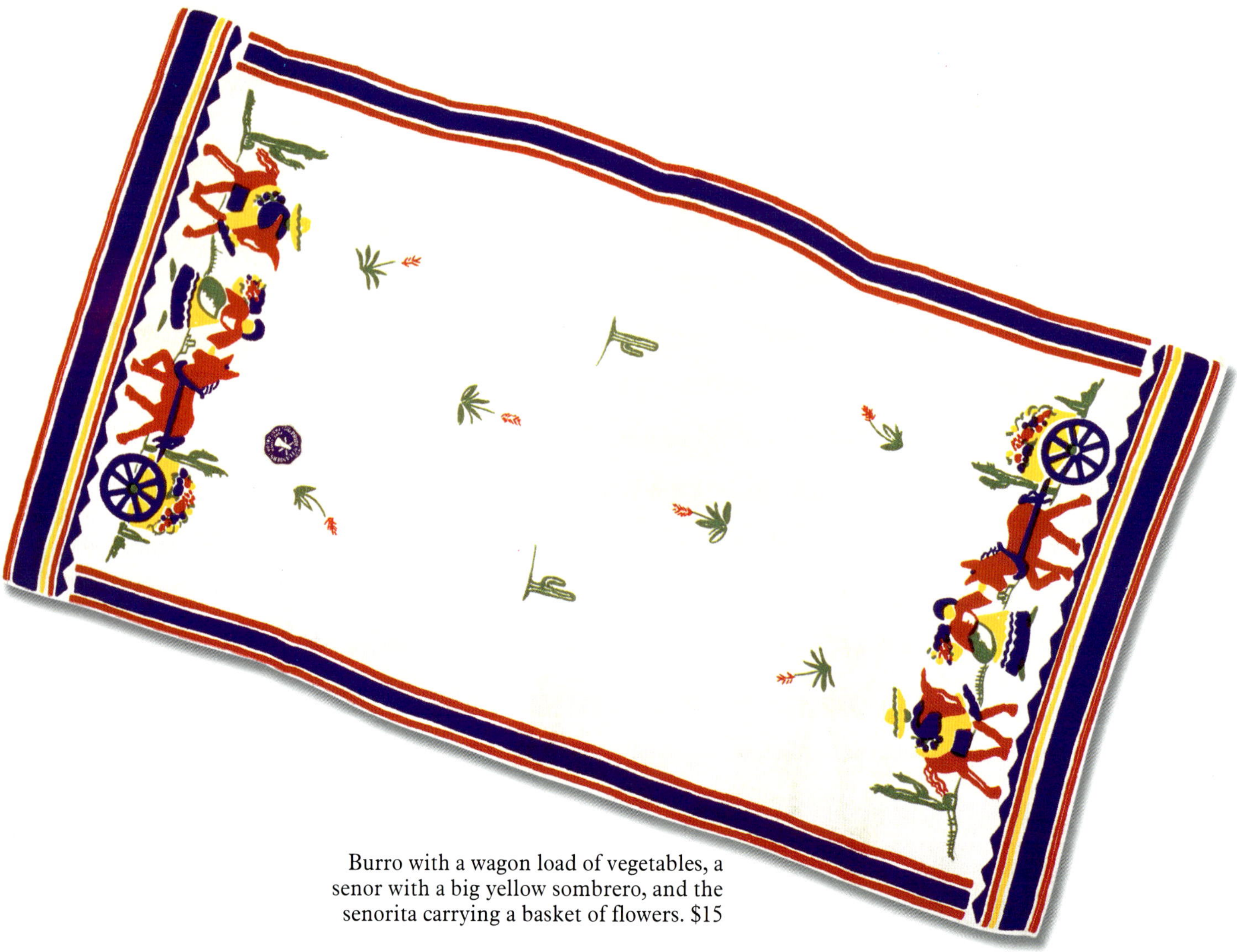

Burro with a wagon load of vegetables, a senor with a big yellow sombrero, and the senorita carrying a basket of flowers. $15

Mixed group towels.

Lobsters and clams. $20

At the marketplace with the gentlemen bartering for their goods of richly colored and striped pottery. $15

Red tulips with yellow on a baby blue grid background. $15

Tea towels in squares and diamonds in orange, blue, green and yellow. Note the fringe which shows a continuation of the colors in the stripes. Set of two. $15

Senor sleeping under a palm tree while his patient burro waits. Red and blue checked baskets full of "cut out" posies. Note the shape of the burros' legs and the flared pants and pointed shoes of the senor. $20

White dogwoods in the **Lakewood** design with hand done tatting in pink on both ends. $15

Striking scene of yellow landscape with gray bricked wall, large colorful urns, and beautiful floral serapes. This is a perfect example of the rich brilliancy of colors Weil & Durrsè are famous for. $20

Brown cat tails and green leaves. $20

Antique key and lock design on blue background. $15

Pineapples, strawberries and grapes with a red border. Artist: John Madsen. Design date: 1939. $15

Row of plums, peaches and pears on each side with a striped border. $15

Linen towel with red strawberries on a tan background. $15

Display of red and yellow cosmos with green border. $15

A background of stripes showing a display of wine jugs, pack of cigarettes and a smoking cigarette in ash tray. $20

Floral display with green border and bright stripes on sides. $15

The Wil-Dri towels were made of soft, 100% cotton with a nubby texture. $15

Simple design of flowers in lemon yellow. $10

Fall display of candy corn and gourds. $20

Pot of baked beans with onions and jug of molasses. $15

Bunches of tomatoes and vegetables on both ends and radishes along the sides. $15

Spring display of yellow daffodils with white and blue hyacinths. Note the light and dark green shading of the leaves. $20

Chapter 3
Aprons

Two groups of red roses in the **Royal Rose** design. $15-20

Salad display with wooden bowl, jar of olives, lemons and dill. One pocket on red cloth. $15-20

Lobsters. This one matches the popular lobster and clam tablecloth. The lobster panel fronts a large pocket. $15-20

Two pockets showing a south western scene. Pockets and apron front are bordered in green rick-rack. $15-20

Wonderful wrought iron bowl of fruit, with cheese and crackers, glasses of liquor, slice of cheese, and more. One pocket on the red cloth. $15-20

Large red roses, rose buds, and gray flowers in background. One pocket at the right hand side of the apron. $15-20

Sweet little terrycloth apron with a nosegay of violets tied with a bow. A pocket on each side. $15-20

Sheer material with a cotton panel showing wild roses. $15-20

This small homemade apron was fashioned from a strawberry tablecloth. $10

Large rose design, this one made from a sheer material with cotton rose front and ruffled all around. $15-20

White apron with red rose border. $15-20

Yellow apron with berries. $15-20

Green apron with three pockets in a southwestern design. $15-20

Coral-colored roses. $15-20

Chapter 4
Handcrafted Items

Strawberry bibs. The lady of the house fashioned these bids out of the **Strawberry** design fabric but apparently found them to pretty to put to use. Set of three $20

Lobster bibs. Set of two $20

Decorative strawberry pillow. $25

Adorable cherry teddy bear. $30

Tote bags. Created from cherry and rose tablecloths, these are beautiful and useful. $30

Blue floral pillow with blue braided trim. $30

Princess rose and cherry pot holders. Set of two potholders $20

Cherry potholder with polka dot fabric for backing. Set of two potholders $20

Small pillow with dainty roses and pink braiding. $30

Red rose pillow with red and white ruffle trim. $30

Pillow with pink roses, pink and white checked sides, and tiny rose ticking fabric for accent.

Chapter 5
Laundering and Storing Linens

Some basic cleaning techniques are good to know.

Stains

First, it is important to realize that many old stains on these cloths are there for good. Not all stains will come out, even with spot treating and soaking. When purchasing a vintage linen, it is best to avoid badly stained items unless you can live happily with them.

After obtaining a stained linen, always give it a thorough rinsing in clean water before laundering it. This will rinse away excessive dirt so that if you plan to soak the cloth, it will not be soaking in dirty water. If there are just a few small stains, I first spot treat, leaving de-stainer on for half an hour. If using a powder de-stainer, continue to add small amounts of hot water to the mixture to keep it a wet paste. After spot treating, soak linen for at least four hours or overnight in your favorite detergent with stain remover. If the cloth has many stains or overall dinginess, skip the spot treating and give it a soak.

Although some stains may disappear quickly others, may not, even after an overnight soak. If stains remain, soak once more overnight in a fresh mixture, then launder.

Hanging linens outdoors on a line in sunshine works well for brightening white linens. Placing damp linens on green grass in sunlight also can remove stains, brighten whites, and refresh old linen. And there is nothing like the clean grassy scent of linens that have been dried outside in the fresh air.

White vinegar is a wonderful natural brightener for dingy or yellowed linens. It also works well for removing musty and mildew odors from fabric that was stored many years.

I do not recommend using chlorine bleach. Over time, it will fade colors. Remember, colorfast does not mean un-fadable.

After soaking, if the item is still stained, I advise simply accepting the stains. I generally do not do more than two overnight soaks on an item because I have found that continually trying more treatments can damage the fabric and cause holes or discoloring. A family probably enjoyed dinner over these cloths 40 or 50 years ago. Instead of fretting over a few spots, I muse on the stories that may have been told over dinner back then and appreciate the history of the cloth. I am pleased to be able to use this same linen in my home all these years later.

Laundering

Using your favorite detergent, launder by hand in the sink or, if the linen is sturdy, launder linens in the washing machine on the delicate cycle only. If laundering in the machine, I let the linens soak for 15 minutes to loosen soil and then set the agitator to run for only 1 or 2 minutes. Longer cycles of agitate can cause damage on a vintage linen.

Rinse thoroughly. Always put linens through two cycles of rinse. Check during the second cycle and if there are still soap suds, rinse again. If laundering by hand, rinse repeatedly until the water runs clean.

If I am going to be using the cloth on the table, I starch it using Argo starch in the final rinse cycle. Spray starch and liquid starches which are used in the rinse cycle work well also, but I prefer the old-fashioned powdered Argo starch as I have been pleased with the crisp, professional look it gives the linen. When drying, whether hanging dry or machine drying, make sure to remove the cloth while it is still slightly damp. Many fabrics iron easier when they are damp.

Laundering Delicate Linens

Any linens that are delicate, worn thin, or have holes should be hand-laundered in the sink or in a wash bucket. Work up a good soapy mixture and gently swish the item through the water, then let it soak, and swish several more times to release any soil. Rinse thoroughly, several times, until the water runs clean. Pat these items dry between towels and lay them flat to dry or line-dry on a calm, not windy, day.

Pressing

I use an Ironrite mangle iron for pressing. Mangle irons are wonderful old machines and are especially appreciated when ironing large tablecloths. They also make a speedy job when pressing a great number of small items. All of my sturdy linens are pressed using my mangle iron, but for the delicate linens I press with a hand iron.

If you do not plan to use a linen immediately after laundering, and instead will be storing it, do not use starch or press the item. Launder, rinse well, and after allowing it to dry thoroughly, gently fold it and store it flat on shelves that are covered with white towels made from 100% cotton.

Do not store your linens directly on painted, varnished, or likewise-treated wood, as chemicals from the coatings can affect the textile, causing discoloration to the fabric.

The Ironrite machine.

Index of Patterns